BIRDWATCHING ON

Whooper Swan

BIRDWATCHING ON MERSEYSIDE

by

Ken Stoba

CICERONE PRESS,
MILNTHORPE, CUMBRIA

ISBN 1-85284-017-X

Front Cover:
Top: Young barn owls *Bottom: Pintail*

CONTENTS

FOREWORD

The Mersey Estuary means many things to many people. It conjures up visions of Liverpool as a great port, of the Beatles, and on the negative side, of pollution. The birds of the Mersey are a hidden treasure, in general known only to keen bird-watchers. The estuaries of the Dee and the Ribble are perhaps better known for their bird life than is the secret Mersey. All three estuaries, to a greater or lesser extent, face threats which could seriously affect the teeming wildfowl and waders that depend upon their mud banks and saltmarshes.

A desire to protect the habitats of birds should spring from an increased awareness of the variety of birds that can be found around our particular localities. Within the following pages you will find many examples of the tremendous opportunities for bird-watching in Merseyside. If this book encourages you to take a greater interest in the birds of Merseyside and, through that interest, develop a desire to see the birds and their habitats protected, then 'Birdwatching on Merseyside' will have played an important role in bird conservation.

Andre R. Farrar
Conservation Officer, Royal Society for the Protection of Birds, North West Region.
November 1988

INTRODUCTION

Primarily undertaken as a brief outline of the main birdwatching sites in a relatively small area, this guide quickly developed both in length and in breadth, stimulated by two principal causes. First, as site followed site, more were recalled or suggested and, having appeared, I could find no good reason for including one to the exclusion of others. Secondly, as I sought information and corroboration from other sources it became apparent that a book of this scope was both needed and wanted, and the enthusiasm and encouragement of other people was such that I was duty bound to do my best to come up to the mark.

Without exception, the people with whom I came into contact, Wardens, Reserve Management, Council Officials, Group Members and Representatives of other Associations as well as private individuals, all gave unstintingly of their time and freely of their advice and knowledge; I extend my sincere thanks to them all. As the number of locations increased the contents were divided into four approximately equal sections corresponding to the rough geographical boundaries heading each. For convenience sake I have assumed that the areas bounded by the River Alt should be divided as though the Alt ran due East from its mouth. Similarly, the inclusion of a large chunk of Cheshire in a book titled "Merseyside" is excused by the excellence of some of the sites just as close to the south of Liverpool as Southport is to the north, begged for their inclusion. I'm sure the good people of Cheshire won't mind this bit of 'poaching' (– some of my best friends are Cestrians).

Ornithologically speaking, the area covered is dominated by the presence of the massive estuarine systems of the Ribble, the Mersey and the Dee. Each has its own character and attraction to birds individually but jointly combines with its neighbours to provide the system which makes a narrative bias towards the nationally and internationally important numbers of waders and wildfowl practically unavoidable. Although its quality as such cannot be overstated, neither the presence of other species nor the value of inland sites is insignificant; far from it. The overall result is an area which produces some of the best birdwatching in Europe, and I jest not!

On visiting even those locations which I knew, or thought I knew, completely, it soon became apparent that 'knowing' and communicating that knowledge were two entirely different things. Paths change, signs are removed or are non-existent, so even my favourite sites had to be re-visited with this in mind and the information verified and up-dated, but the very nature of a guide book means that information and directions given are likely to suffer variations. The information given is accurate at the time of compilation, but the author would be glad to receive any information concerning changes. Where the names of public houses had to be checked this task became less onerous!

All in all some 58 sites are included and, on counting back, an astounding 214 species of birds have been mentioned – not a comprehensive list for the area by any means, as a scan through the Lancashire and Cheshire Bird Reports will show. It has, obviously, not been possible to include all the locations in which birds are seen, nor all the birds to be seen in each area, so, please accept my apologies if your favourite site or special bird doesn't appear. I do hope however that those included are amongst the best, but remember – anything can turn up anywhere – like recent records of a Little Auk in the road near Scarisbrick or a Stork on a school roof in Woolton.

This book must inevitably be dedicated to the stars of the show – the birds! Good watching!

Ken Stoba

Red Necked Grebe

ABBREVIATIONS & SYMBOLS USED IN THIS BOOK

B.S.C.	British Steel Corporation
C.C.C.	Cheshire County Council
L.N.R.	Local Nature Reserve (see Local Council)
L.T.N.C.	Lancashire Trust for Nature Conservation
L.U.W.U.	"Landlife" Urban Wildlife Unit
M.B.C.	Metropolitan Borough Council
M.V.P.	Mersey Valley Partnership
M.N.A.	Merseyside Naturalists' Association
N.C.C.	Nature Conservancy Council
N.N.R.	National Nature Reserve (N.C.C.)
N.T.	National Trust
R.S.P.B.	Royal Society for the Protection of Birds
S.S.I.	Site of Scientific Interest (N.C.C.)
S.S.S.I.	Site of Special Scientific Interest (N.C.C.)
W.E.C.G.	Woolston Eyes Conservation Group
W.F.T.	Wildfowl Trust

KEY TO SYMBOLS & LOCATION HEADINGS

® – Reserve [(W) = Permanent staff see appendix for Warden/Manager's Details under (W.F.T.), (R.S.P.B.) etc.]

Ø – Opening/access restricted

(W.C.) – Toilets

© – Admission/parking Charge

☛ – Public access

(Dis) – Disabled Access toilets()

(P) – Permit required

⇌ – Railway Station within reasonable distance, name in brackets ()

(Café) – Café/Refreshments available.

References following 'LOCATION' details are full Map References referring to Ordnance Survey 'Landranger' Maps commencing with Sheet Number in Brackets ().

Bewick's Swan in flight. (Shotwick Lake, see p.85)

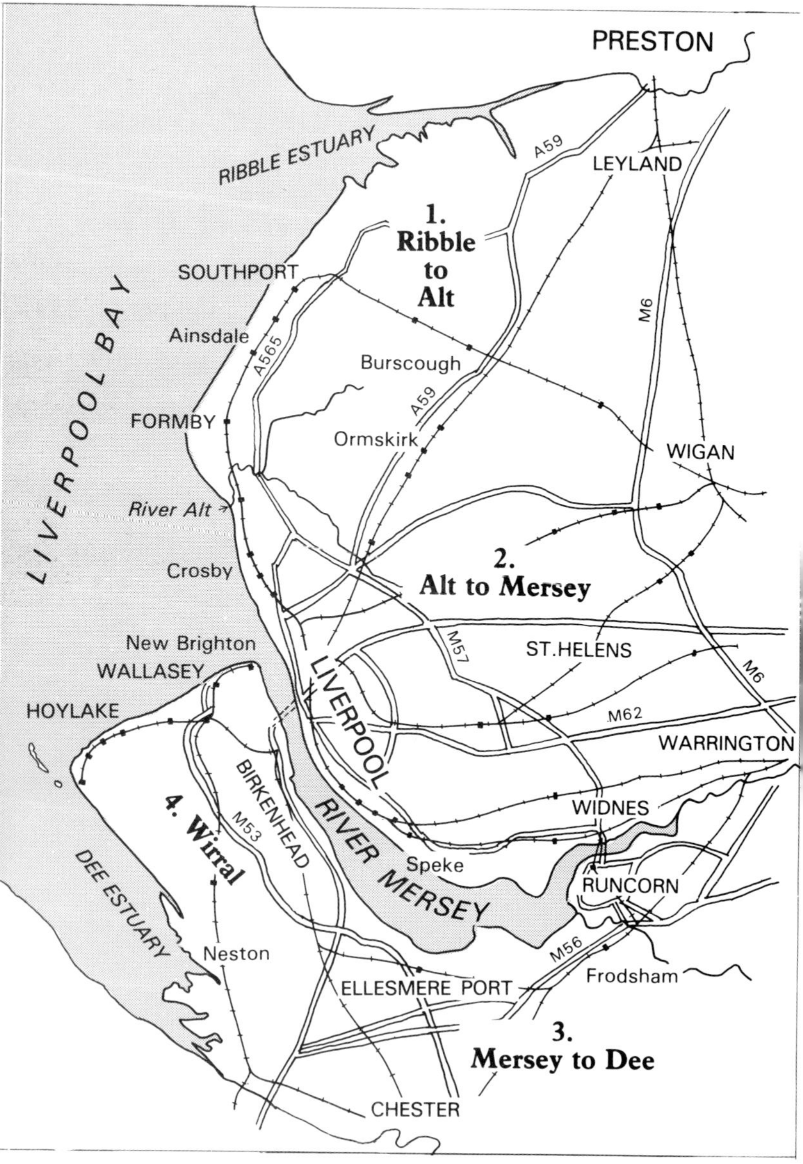
PRESTON
RIBBLE ESTUARY
A59
LEYLAND
1. Ribble to Alt
SOUTHPORT
LIVERPOOL BAY
Ainsdale
A565
Burscough
M6
A59
FORMBY
Ormskirk
WIGAN
River Alt
2. Alt to Mersey
Crosby
New Brighton
M57
ST. HELENS
WALLASEY
LIVERPOOL
M6
HOYLAKE
M62
WARRINGTON
BIRKENHEAD
4. Wirral
M53
RIVER MERSEY
WIDNES
DEE ESTUARY
Speke
RUNCORN
Neston
M56
Frodsham
ELLESMERE PORT
3. Mersey to Dee
CHESTER

Lesser Black Backed Gull

SECTION 1.
RIBBLE
to
ALT

HESKETH MARSHES

☛ Footpaths, Minor Roads, Ⓡ N.C.C. (part), ⇌ (Crossens).

LOCATION
South of River Ribble between Preston and Southport (102) SD 454209

ACCESS
From A59 at Tarleton take Coe Lane opposite small church, signposted Hesketh Bank/Tarleton and follow road left around bend, turning right into Hesketh Lane.

For route 1. turn right into Sutton Ave. (3rd road after school), parking on waste ground at bottom of road, follow path to left along river bank. Leave embankment by way of the paths* to the left and return by road.

For route 2. continue on Hesketh Lane down the 'T' junction signed left Southport, right Private Road. Turn left, then first right, parking in Guide Road. The Private Road is a public footpath which continues left at Marsh Farm to the embankment path*. Alternatively park further along Guide Road gaining access to embankment by paths on right.

HABITAT
Riparian grassland with arable fields, some broad-leafed trees and hedgerows on inner marsh and riverbank; progressing to grazing land, saltmarsh and estuarine mud and sand on outer marsh.

BIRDS
A good variety of farmland and estuarine species breed including **Corn Bunting**, **Sedge Warbler**, **Oystercatcher**; **Spotted Redshank** may be seen among the many other waders which occur on migration. **Hen Harriers**, **Short-eared Owls** and **Pink-footed Geese** are winter visitors. Rarities and occasional visitors include **Marsh Harrier**, **Brent**, **Bean** and **White-fronted Geese**.

Since the reclamation of the outer marsh* this area is now underwatched but the variety of routes make an interesting summer walk and it is well worth including, at least, one of the shorter routes to the outer embankment in any winter visit to the area.

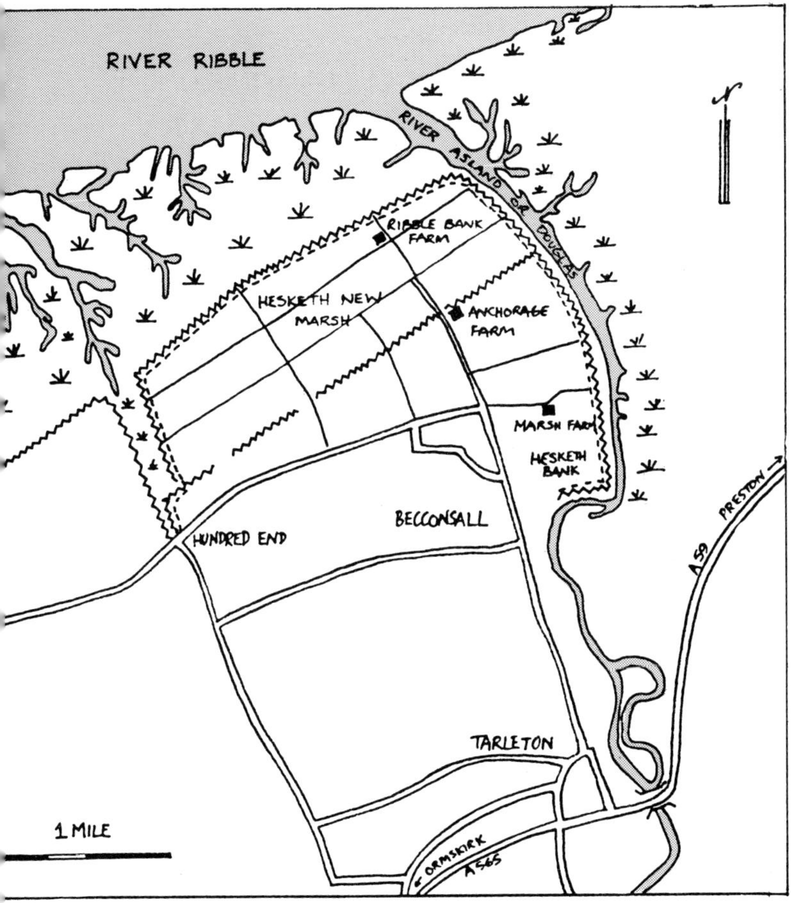

HESKETH MARSH

*NOTE – Public access by vehicle is only as far as the end of the tarmac just past Anchorage Farm. The remaining saltmarsh at the Hundred End end of the embankment is part of the Ribble Marshes N.N.R. and should not be entered without a Nature Conservancy Council permit.

The unmetalled roads between the inner embankment at Ribble Bank Farm and the outer embankment are *NOT* public footpaths; please keep to the embankments.

SOUTH RIBBLE MARSHES N.N.R. & SOUTHPORT SANCTUARY
(including Marshside and Inner Crossens Marsh)

☛ Roads, Footpaths, Shoreline. Dis. ⇄ (Southport). ®(W)[(P) part]

LOCATION
South Ribble Estuary North of Southport (102) SD 353206.

ACCESS
1. Inner and outer Crossens Marsh: From Coastal Road embankment running Northwest from Southport to A565(T) roundabout at Banks. The shoreline at outer Crossens may be followed from the left-hand side of Marshside sandworks at the end of Hesketh Road.*
2. Marshside and Inner Crossens Marsh: From Hesketh Road and Marshside Road.
3. Banks Marsh**: From roundabout at Banks via Banks Rd.
   (i) Follow embankment path on left from opposite Banks Pumping Station.
   (ii) Follow embankment path from Hundred End.

HABITAT
Estuarine mud and sand, saltmarsh, coastal dunes and scrub, rough pasture with scattered trees.

BIRDS
This area contains some of the best sites in the whole country for observing migrating and over-wintering waders and wildfowl. Migration in spring and autumn brings in **Knot**, **Dunlin**, **Bar-tailed** and **Black-tailed Godwits**, **Oystercatcher**, **Grey** and **Ringed Plovers** in flocks, which in the case of the **Dunlin** can be in five figures, to re-fuel on the food-rich mudflats. Considerable numbers of these birds stay for the winter to be joined by the many wildfowl which come to bolster the resident population of **Mallard** and **Shelduck** roosting on the mudflats and feeding on the saltmarsh. The Southport area in general is well-known for the substantial quantities of **Pink-footed Geese** to which it plays host and which frequent these marshes along

Right: Goldfinch

Above: Purple Sandpiper

Right: Grey Lag Geese in flight

with **White-fronted**, **Brent**, **Bean** or even the odd **Snow Goose**, all of which occur with reasonable regularity. Ducks are well-represented with **Wigeon, Teal, Shoveler, Pintail** and sometimes **Long-tailed Duck** or the rare **Blue-winged Teal**.

The 'common' **Mute Swan** is eclipsed by the large herds of **Whooper** and **Bewick's Swans** which spend the winter here, tending to favour Outer and Inner Crossens respectively, both species have built up into three figure numbers over the last few years.

Although waders and wildfowl provide the bulk of the attractions, other migrant and winter-visiting species can also be of considerable interest. Small flocks of **Twite** and **Snow Buntings** are often seen with the commoner **Greenfinches** and **Linnets** along the embankments; **Merlin** and **Peregrine** find a plentiful food supply as do the healthy numbers of **Short-eared Owls** which patrol the marshes and grazing land.

When summer arrives the frantic activity of waders and wildfowl may have ceased but several species use the area to breed. **Black-headed Gulls**, **Redshank**, **Common Terns** and **Shelduck** favour the out-marsh whilst **Linnets**, **Meadow Pipits** and **Skylarks** extend into the farmland.

GENERAL

The Reserve marshes are managed by controlled summer grazing which maintains the grass height at acceptable levels for both breeding birds in summer and grazing birds in winter.

Shooting takes place in some areas between 1st September and 20th February, except on Sundays.

N.N.R. Area

Permits are required for access to the study zone and Hesketh Out Marsh. Organised groups should apply to the warden in advance giving proposed date, number in group and purpose of visit.

*NOTE: You should not attempt to walk across or around the Marsh without first checking the times of tides, some tides cover the Marsh, also the gulleys are deep and fill quickly.

** Cars are *not* allowed down any of the private roads and tracks leading to the embankment.

Left top: Greenfinch Below: Snow Bunting

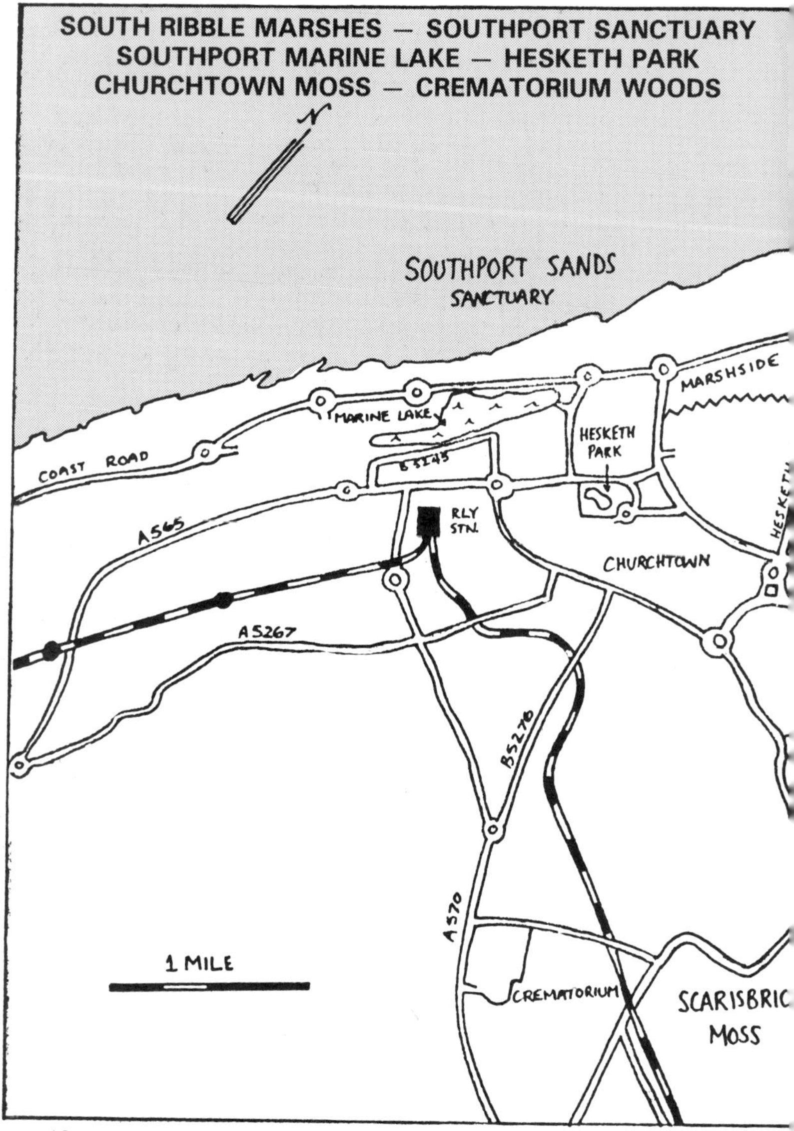
SOUTH RIBBLE MARSHES — SOUTHPORT SANCTUARY
SOUTHPORT MARINE LAKE — HESKETH PARK
CHURCHTOWN MOSS — CREMATORIUM WOODS
N
SOUTHPORT SANDS
SANCTUARY
MARSHSIDE
MARINE LAKE
HESKETH PARK
COAST ROAD
B5245
RLY STN.
A565
CHURCHTOWN
A5267
B5276
A570
1 MILE
CREMATORIUM
SCARISBRIC
MOSS

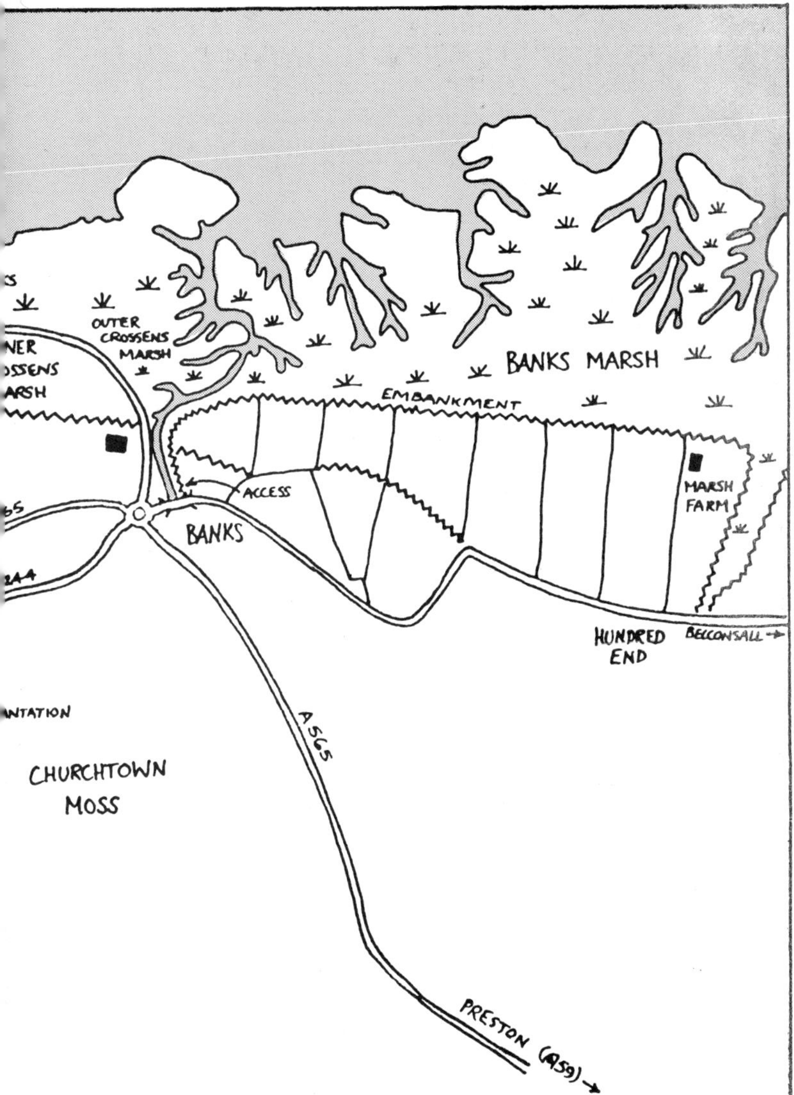
OUTER
CROSSENS
MARSH
BANKS MARSH
EMBANKMENT
ACCESS
BANKS
MARSH
FARM
HUNDRED
END
BECCONSALL
A565
CHURCHTOWN
MOSS
PRESTON (A59)

MARTIN MERE WILDFOWL TRUST

☛ ® (W.F.T.) Ø © W.C. Dis. (T)* ⇌ (Burscough), Café.

LOCATION
5m. East of Southport. (108) SD 430144.

ACCESS
From A565(T) Southport/Preston road via B5246 at Mere Brow following Wildfowl Trust signs.

OR
From A59(T) Ormskirk/Preston road following signs along Red Cat Lane at railway bridge in Burscough Bridge town centre.

HABITAT
Managed reserve area with mere, peat mossland, artificial lakes and wader scrapes, ecologically planted shrubs and trees.

BIRDS
One of the best bird-watching sites in the region; the reserve, whilst having an interesting variety of terrestrial birds like **Corn Bunting**, **Pheasant**, **Grey Partridge** and **Barn Owl** is, quite rightly, of prime importance for its waterfowl. **Mallard**, **Shelduck**, **Gadwall**, and **Redshank** breed as do **Coot** and **Moorhen**, but it is winter that provides the greatest draw when the "wild" area attracts vast quantities of winter wildfowl including 200+ **Whooper** and 400+ **Bewick Swans** with varying numbers of **Pintail**, **Teal**, **Wigeon** and **Shoveler**. There is a spectacular flock of **Pinkfooted Geese** which regularly reaches 7,000 birds – best seen in the evening when the birds, which may have been absent during the day, return to roost. These wildfowl are augmented by visiting **Greylag Geese** and a flock of feral **Barnacle Geese** with the odd genuine **Barnacle**, **Bean** or **Whitefronted Goose**. Waders include **Ruff** (which are present all year and have bred on one occasion, building up to the largest winter flock in the country with up to 500 birds present), **Snipe**, breeding **Little-ringed Plover**, and migrating **Little Stints**, **Spotted Redshank** and **Greenshank**.

Many birds of prey hunt the area with **Sparrowhawk** and **Kestrel**

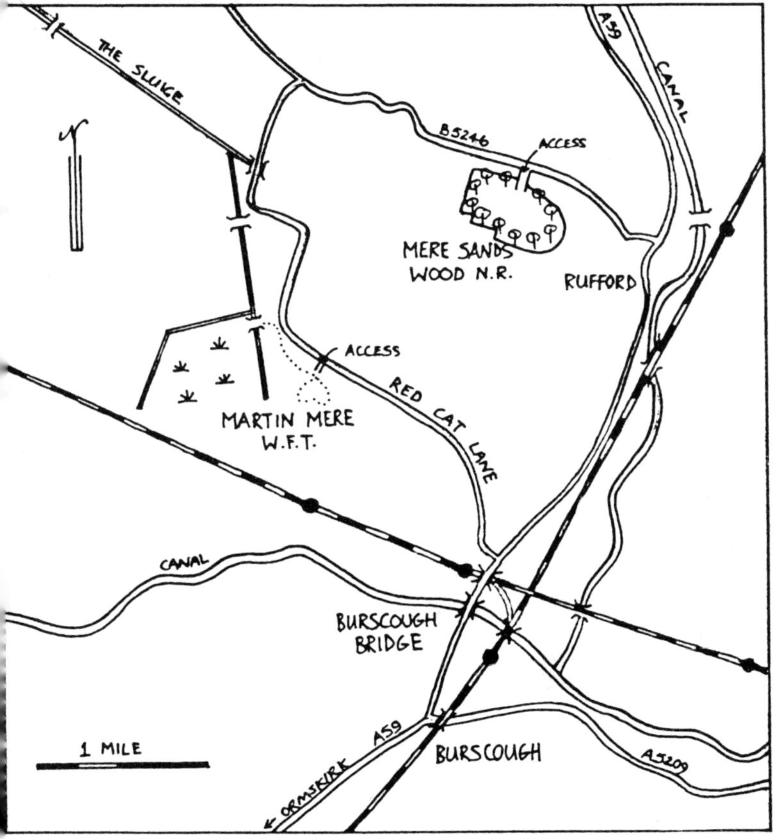

MARTIN MERE WILDFOWL TRUST
MERE SANDS WOOD

being the commonest. **Short-eared Owls** are present in the winter as are visiting **Hen-Harriers** and **Merlin**. **Marsh Harriers** occur as summer visitors with the occasional rarity like **Hobby** or **Osprey**.

The **Stock** and **Collared Doves** breed locally as the reserve plays host to many of our commoner summer visitors. One area has been especially designed in the hope of inducing visiting **Kingfishers** to breed.

Ruff

The overall attractiveness of the area to birds, combined with the large number of watchers, means that rarities are regularly recorded – the list is too long to mention but includes species like **Spoonbill**, **Franklin's Gull**, **Ring-billed Gull** and **Collared Pratincole**.

GENERAL

Man-made embankments surround the "wild" area, protecting the birds from disturbance, and observers use the well-appointed permanent hides which are built into them. The remainder of the grounds are taken up in part by the Visitor/Information Centre, Café and Shop with the rest being devoted to the "collection" area – open landscaped pens which contain captive wildfowl from around the world. The Trust has a well-developed captive-breeding programme to help further its work in protecting and preserving the world's wildfowl.

Disabled persons are well-catered for and facilities include tape-guides and Braille signs for the visually handicapped.

MERE SANDS WOOD N.R.

☛ Nature trail, Public Hides W.C. ® (L.T.N.C.) (W) ⇄ (Rufford)

LOCATION
1 m. West of Rufford. (108) SD 448162

ACCESS
From A59(T) Liverpool/Preston road via B5246 at Rufford or A565(T) Southport/Preston road via B5246 at Mere Brow the reserve is situated on the south side of the road between Holmeswood and Rufford and car parking is at the end of the track signposted from the road.

HABITAT
An attractive collection of habitats including lakes, heath, broad-leaved and coniferous woodland all brought under management and with the addition of a controlled level wader scrape and shell bank.

BIRDS
The cross-section of available habitat ensures that a considerable

Ruddy Duck (m)

number of species breed; including **Little Ringed Plover**, **Ruddy Duck, Dabchick, Linnet, Treecreeper, Long-tailed** and **Willow Tits.** With about five pairs usually present, the reserve is an excellent place to observe, in March and early April, the elaborate courtship displays of **Great Crested Grebes** – a water ballet of unpaling beauty. The wader scrape attracts many migrants such as **Spotted Redshank** and **Greenshank**, whilst to the lakes and woodland come winter wildfowl and occasional visitors like **Lesser-spotted Woodpecker** and **Kingfisher**. In all, a good place to visit at any season, which, with continued good management, can only get better.

GENERAL

The reserve covers an area of about 100 acres and is owned and managed by the Lancashire Trust for Nature Conservation. Several well-appointed public hides and a way-marked nature trail are a substantial bonus as is the added interest of the local plant and animal life.

Ringed Plover

S.W. LANCASHIRE MOSSES

☛ Public footpaths and minor roads, Dis.

LOCATION

A massive area of drained marshland stretching up to six miles inland from Southport in the north to Hightown in the south and divided for our purposes into three areas, namely:–

1. Churchtown Moss (108) SD 374177 **(See map p.18)**
2. Halsall and Plex Mosses (108) SD 368100
3. Downholland and Altcar Mosses (108) SD 361081

These divisions are for convenience of narrative only – the areas are adjoining and distinct separation on purely ornithological grounds is impractical.

HABITAT

Totally flat, abundantly fertile mossland and fine sandy soil converted to arable and grazing land, crossed by drainage ditches and hedgerows with scattered patches of deciduous woodland.

ACCESS

1. At A5267 Churchtown roundabout take Moss Lane (2nd exit when approaching from south) and follow road to right at a plantation, then signs to Bescar. On reaching Bescar Lane return may be made either to the left via minor roads or right to B5242 and thence A50(T).
2. (i) From the A567/A5267 Southport/Maghull Road to the east: via B5243 Jacksmere Lane (Scarisbrick), Gregory Lane or Carr Moss Lane (Halsall).
   (ii) From the A565(T)/A570(T) to the west: via B5243 Moss Lane or Halsall Rd. (Birkdale), Segars Lane (Ainsdale).
3. (i) From the A567/A5267 Southport/Maghull Road to the east: via Blue Bell Lane or School Lane (Haskayne), B5195 (Downholland).
   (ii) From the A565(T)/A570(T) to the west: Moor Lane (Ainsdale), B5195 Moss Side (Formby).

The many other minor roads and footpaths crossing the area between those mentioned above provide plenty of scope and are well worth further investigation.

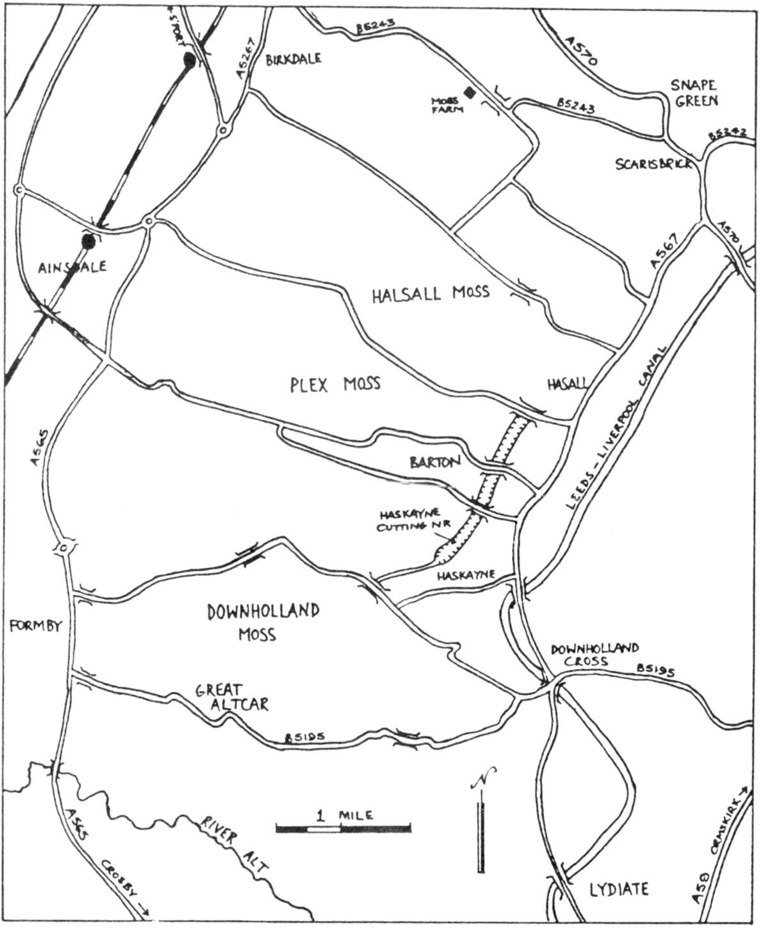

S.W.LANCASHIRE MOSSES
HASKAYNE RAILWAY CUTTING N.R.

N.B. Several of the above minor roads have a 2 ton weight restriction and most are unsuitable for coaches.

BIRDS

A wealth of birds throughout the year provide good prospects for any visit. Regulars include **Corn Buntings**, **Yellowhammers**, **Turtle Doves**, **Red-legged** and **Grey Partridges**. Primarily known as a stronghold for the endangered **Barn Owl**, the area is also hunted over by **Little**, **Tawny** and **Long-eared Owls** with the rare **Montagu's Harrier** and **Rough-legged Buzzard** having been recorded by day. **Dotterel** are an annual spring migrant although somewhat hard to locate, **Woodcock** breed in the woodlands and **Water Rails** occupy the ditches. Despite all of these, the main claim to fame of the area is the attraction which it holds as a feeding ground for winter geese. In fact, an internationally significant percentage of the world's population of **Pink-footed Geese** migrate here; keeping company with the odd **White-fronted**, **Bean** or **Snow Goose**. The geese move around the area feeding mainly in the stubble fields, so a little while spent searching is worthwhile in order to guarantee results.

Please take particular care not to disturb the geese unnecessarily. A large proportion of the area is shot-over, quite a lot of it being managed as pheasant shooting, so the geese are quite naturally nervous of any approach made on foot. The best views are really obtained from the car.

The bridge on the B5243 nr. Moss Farm and the bridges on School Lane, Moss Side (Haskayne/Formby) rise above the flat mossland and make excellent vantage points for locating the winter geese.

SOUTHPORT MARINE LAKE

☛ Dis. W.C. ⇄ (Southport), Café.

LOCATION
Southport (108) SD 338180

ACCESS
B5245 Marine Drive.

HABITAT
Artificial marine lake and islands backed by dunes.

BIRDS
Migrant and winter waders and wildfowl, with good numbers of **Goldeneye** regularly recorded along with **Red-Breasted Merganser** and the occasional **Scaup**, **Common Scoter** or **Greylag Goose**. Gulls include **Common**, **Greater** and **Lesser Black Backed**, more rarely **Little**, **Glaucous** and even **Bonapartes** on one occasion.

A good place to combine with an out of season visit to any of the other nearby locations but summertime recreational use disturbs all but the most stoical residents at that time. Broadly speaking, the worse the weather the better the prospects.

AINSDALE N.N.R. & FISHERMAN'S PATH

⇌ Footpaths, ® (N.C.C.) (W) [(P) part], ⇌ (Freshfield).

LOCATION
Lancashire coast between Formby and Ainsdale (108) SD 292092.

ACCESS
From A565(T)/B5424 roundabout (Formby North) exit via Southport Rd. (B5424). Follow the road around two sharp bends and straight on into Piercefield Road, then Old Town Lane, turning right at 'T' junction into Freshfield Road, then left into Victoria Road. Park in Freshfield Station car park. Access to the reserve is along the footpath next to the railway, crossing the line at the un-manned level crossing, ½ mile north of the station.

HABITAT
Pinewoods, fixed and mobile dunes, foreshore.

BIRDS
The un-thinned pinewoods are predictably sterile but the clearings and periphery of the waymarked woodland paths encourage a good selection of woodland birds such as **Greater Spotted Woodpecker**, **Sparrowhawk** and **Tawny Owl**. In summer the smaller Passerines include **Willow** and **Grasshopper Warbler**, **Whitethroat** and **Chiff-Chaff** joining the resident **Pheasants**, **Woodpigeons** and **Wrens**. In winter the silence of the woods can suddenly be broken by the sibilant calls of large mixed flocks of **Blue Tits**, **Great Tits**, **Coal Tits** and **Long-tailed Tits**, which can appear from nowhere to surround the observer. **Lapwings** and **Meadow Pippets** breed in the surrounding field, which are visited on migration and in winter by **Golden Plover**, **Fieldfare** and **Redwings**. The latter two species taking advantage of the plentiful supply of berries on the Sea Buckthorn. The Fisherman's Path, which is prominently waymarked, passes through the southern end of the woodland before entering an area of fixed dunes, home of **Skylarks**, **Pipits** and **Grey Partridge** and around whose slacks **Shelduck** breed. The path continues through mobile dunes to a large expanse of foreshore with shallow low tide pools,

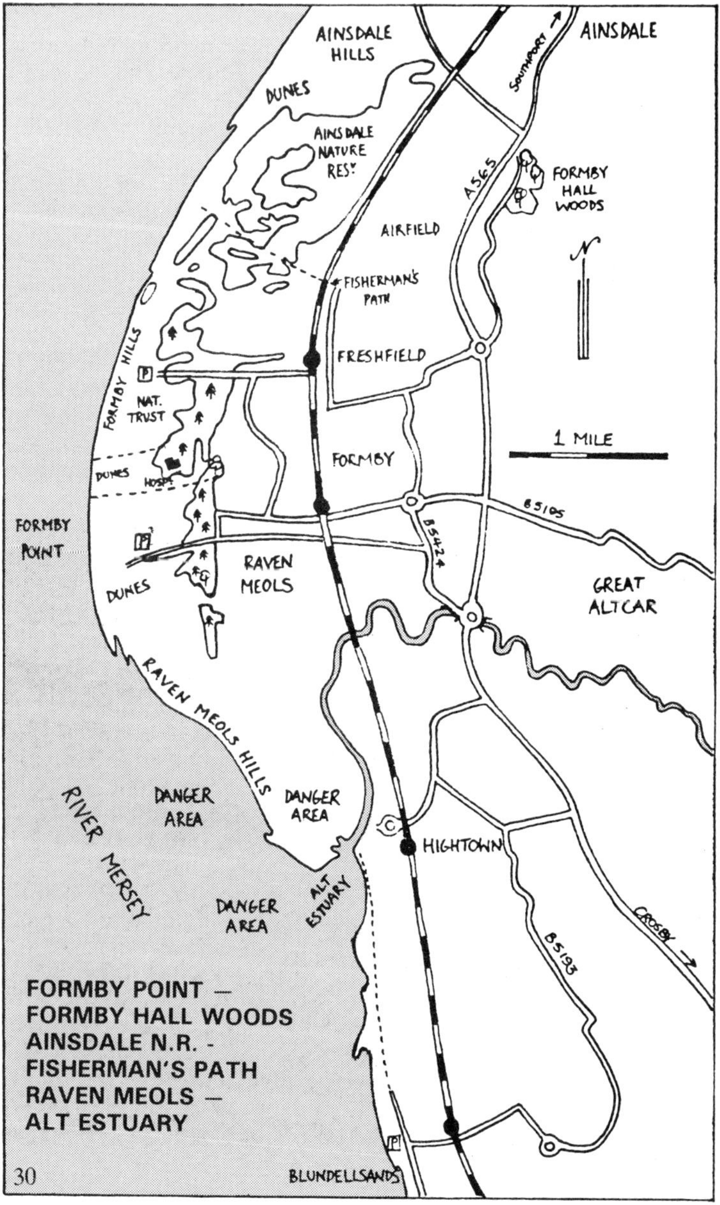

FORMBY POINT –
FORMBY HALL WOODS
AINSDALE N.R. -
FISHERMAN'S PATH
RAVEN MEOLS –
ALT ESTUARY

which is frequented by several species of **Gull** throughout the year and many waders, notably **Sanderling**, **Knot** and **Dunlin** on passage and in winter.

GENERAL

The reserve, run and wardened by the N.C.C., provides a sanctuary for a number of unusual species of insects, plants and fungi as well as the rare and protected Natterjack Toad and Sand Lizard.

Sanderling in winter

FORMBY POINT N.T.

☛ Private Rd. (© vehicles) footpaths; Dis.; ® (N.T.) (W); ⇌ (Freshfield).

LOCATION
Formby (108) SD 281083.

ACCESS
From A565(T)/B5424 roundabout at the north end of Formby, exit via Southport Road. Follow road around two sharp bends straight on into Piercefield Road and then Old Town Lane, turning right at 'T' junction into Freshfield Road then left into Victoria Road.

HABITAT
Pinewoods, willow/birch shrub, fixed and mobile dunes, foreshore.

BIRDS
The changing habitat provides a niche for a wide cross section of species throughout the year; **Greater-spotted Woodpeckers**, **Spotted Flycatchers**, **Willow Tits**, **Willow Warblers** and **Tawny Owls** in the woods; **Redpoll**, **Stonechats**, **Whinchats** and **Yellowhammers** in the scrub and dunes; **Ringed Plover**, **Oystercatchers** and **Gulls** on the beach; the variety of species increasing with the influx of migrants like **Wheatear** and **Ring Ousel**, and winter waders like **Bar-tailed Godwit** and **Dunlin**. Of particular interest is the annual winter gathering offshore of **Common Scoter** and a few **Velvet Scoter** which seem to favour this particular stretch of coast – the former often in appreciable numbers, well into three figures.

Immediately opposite the entrance is an area held as a 'Squirrel Reserve' where Red Squirrels may readily be seen and fed along with the local small bird population who are quite used to human presence.

GENERAL
The area is managed as a reserve by the National Trust and cars are charged for admission.

Right. Shag

Above: Great Spotted Woodpecker
Right Top: Chaffinch (female) *Right Below: Barnacle Goose*

RAVEN MEOLS L.N.R.

☛ Footpaths Shoreline; Ⓡ (Sefton M.B.C.); ⇄ (Formby).

LOCATION
Formby (108) SD 274065.

ACCESS
From A565(T) Southport/Liverpool Rd. Formby roundabout (south) take B5424 Liverpool Rd. then Raven Meols Lane (5th on left) at thatched house. Continue over the railway into Queens Rd. then straight on to car park on right-hand side of Lifeboat Rd.

HABITAT
Fixed dunes, slacks and coastal scrub.

BIRDS
Though not very large in extent this small local nature reserve plays host to a surprising number of species, including breeding **Yellowhammers, Stonechats, Whitethroats, Grey Partridges** and **Linnets. Greater-Spotted Woodpeckers** can often be seen commuting between the adjacent patches of woodland which hold the **Sparrowhawks** and **Tawny Owls** which hunt in the vicinity. Besides the resident **Gulls** the shore attracts migrant and wintering waders like **Knot**, **Dunlin** and **Sanderling**, and rarities like **Shorelarks** have been recorded in the area.

GENERAL
Two rare and protected species, namely the Dune Heleborine Orchid and the Natterjack Toad occur on the reserve.

NOTE: Please observe the Country Code and do not pick wild flowers.

Left top: Fieldfare   Below: Yellowhammer

HASKAYNE RAILWAY CUTTING N.R.

☛ Nature trail, ® (L.T.N.C.), (P).

LOCATION
Haskayne nr. Ormskirk (108) SD 358089.

ACCESS
From A567 Maghull/Southport road via Station Rd. at Bluebell Inn, Haskayne. Park at reserve entrance next to old railway bridge.

HABITAT
A disused railway cutting with rocky outcrops, willow and birch trees, scrub, grassland and marsh.

BIRDS
Nearly a mile of pathway along the undisturbed railway cutting provides access to a good habitat for many farm and scrubland birds with about sixty species having been recorded over the years. Breeding birds include residents like the **Yellowhammer**, **Song**

Song Thrush

Thrush and **Greenfinch**, as well as summer visitors like **Willow Warbler**, **Whitethroat** and **Whinchat**. In spring and autumn migrants seek temporary refuge amongst the willow and birch, whereas in winter, visitors like **Fieldfare** stay longer, sometimes roosting in these same trees. Large flocks of **Pink-footed Geese** visit the adjacent mosslands in great numbers to feed on the winter stubble fields which may be seen extending from the west of the reserve.

GENERAL

The reserve is managed, under licence from the North West Water Authority, by the Lancashire Trust for Nature Conservation, whose volunteers help to maintain the variety of habitat by controlling the invading scrub. L.T.N.C. produces a trail guide which, in addition to acting as a visitor's permit, outlines points of interest. The cutting contains a varied plant and animal life with much to offer besides purely ornithological interest.

HESKETH PARK

☛ Public Park, W.C. Dis., ⇌ (Southport).

LOCATION
Southport town (108) SD 347182.

ACCESS
From Lord Street or Cambridge Road (A565 Preston road) Southport. Parking alongside Park in Albert Road, Park Crescent or Park Road.

HABITAT
Parkland, woodland and shrubbery with artificial lake.

BIRDS
A pleasant and compact location for seeing a selection of the commoner parkland birds, several of the resident species, including **Willow Tits**, **Redpoll** and **Treecreeper** breed, and the woodland is used in season as a temporary refuge by a variety of passage migrants like **Blackcap** and **Garden Warblers**. In winter the resident **Mallards** on the lake are joined by many more of their wild relatives, but not to the exclusion of other wildfowl which also find the lake attractive at this time of the year.

Despite being an 'urban' site its proximity to the nearby bird-rich areas means that a surprising number of less common species may be recorded; the likes of **Merlin**, **Scaup**, **Long-Tailed Duck**, **Waxwing** and **Pied Fly-catcher** for instance and even **Night Heron** on one occasion.

Pied Flycatcher

SOUTHPORT CREMATORIUM WOODS

☛ Footpaths, W.C. Dis.

LOCATION
Between Scarisbrick and Southport (108) SD 365146.

ACCESS
Directly from A570 Southport/Ormskirk Road. Signposted west side of road next to the hospital.

HABITAT
Mature deciduous woodland surrounded by farmland.

BIRDS
A good woodland site which is practically undisturbed with the resultant variety of breeding species like **Spotted Fly Catcher**, **Greater Spotted Woodpecker** and **Willow Tit**. A few of the scarcer species including **Green** and **Lesser Spotted Woodpecker** have also been recorded. The farmland is good for birds such as **Lapwings**, **Skylarks** and **Yellowhammers** and in winter is criss-crossed by many of the wildfowl and other winter visitors more commonly sought after at the better known locations. Under-watched, the wood holds potential as a change from the 'usual' spots.

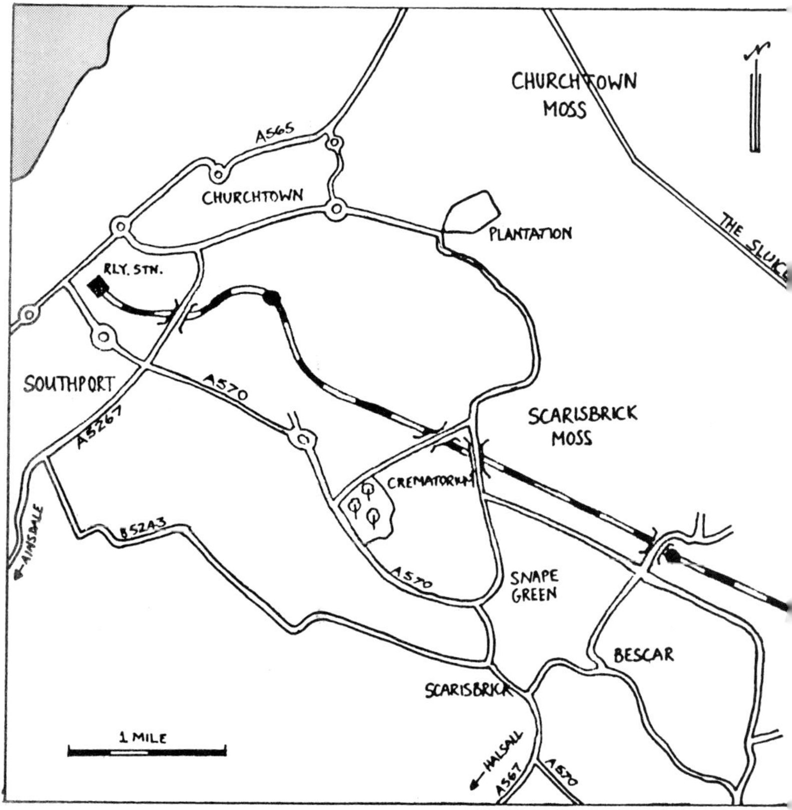

SOUTHPORT CREMATORIUM WOODS
CHURCHTOWN MOSS

FORMBY HALL WOODS

☛ Footpaths.

LOCATION
North of Formby (108) SD 311102.

ACCESS
From A565(T) at Formby roundabout (North), exit via North Moss Lane, turning at the first left into Old Southport Road and following this for about a mile to the Hall.

OR
From A565(T) from Southport, turn left into woods at Old Southport Road, just before start of dual carriageway at Woodvale.

HABITAT
Mature woodland bordered by arable farmland.

BIRDS
Bullfinch, **Treecreeper** and **Goldcrest** are amongst the many woodland species to be found breeding in this fairly small but attractive wood which lies on the edge of the S.W. Lancashire mosslands, thus making it well-worth investigating when visiting any of the better-known sites locally. Records of **Great Grey Shrike**, **Firecrest**, **Goshawk** and **Buzzard** (rarely seen in this area), indicate that a little time spent here can prove to be extremely rewarding.

GENERAL
The grounds and car park of the Tree Tops Restaurant, which is situated in the wood, are private and for patrons only.

Curlew Sandpiper

SECTION 2
ALT to MERSEY

ALT ESTUARY

☛ Footpath, common land, ⇄ (Hightown).

LOCATION
Hightown south of Formby (108) SD 296037.

ACCESS
From A565 Southport/Liverpool Road take B5193 Orrell Hill Lane following signs for Hightown to roundabout. Exit via Lower Alt Road (third exit), parking at end of road. Follow footpath left along river bank. North bank access from Formby along foreshore is restricted when Altcar rifle range is in use; a red flag is flown.

Alternatively, the estuary can be reached after a walk of about two miles along the coast from the seafront car park at Hall Road West/Burbo Bank Road North, Blundellsands.

HABITAT
Small reed bed, dunes, estuarine mud and sand, coastal scrub.

BIRDS
Variety throughout the year with **Stonechat**, **Whinchat**, **Grasshopper Warbler** breeding; a good migrant passage including **Arctic**, **Common** and **Little Terns** with the occasional **Black Tern** or **Stint**. Winter waders include **Dunlin**, **Turnstone**, **Godwits** and **Sanderling**. Perhaps the most notable feature is the large local population of **Knot**, often seen together in impressive numbers performing their closely co-ordinated wheeling and turning flight, flashing white then grey as the flock banks and veers in the winter sun. **Little** and **Common Gulls** are regular in winter with sometimes a few **Snow Bunting** and **Twite**.

Outside the breeding season, **Red-breasted Mergansers** can often be seen fishing for sand-dabs in the channel.

A good spot for rarities, which have included **Red-necked Phalarope**, **Hobby** and **Hoopoe**, especially in autumn.

Easy access from the south side and the comparatively concentrated area make for easy watching, especially in the morning when the light is behind. The coastal vegetation provides a frequently used refuge for migrants and, as the approaching tide pushes the shore-birds closer to the observer, the flat tops of the outfall pipes provide excellent vantage points.

GENERAL

This area lies at the southern end of the largest continuous stretch of sand-dunes in the country, extending northwards as far as Southport.

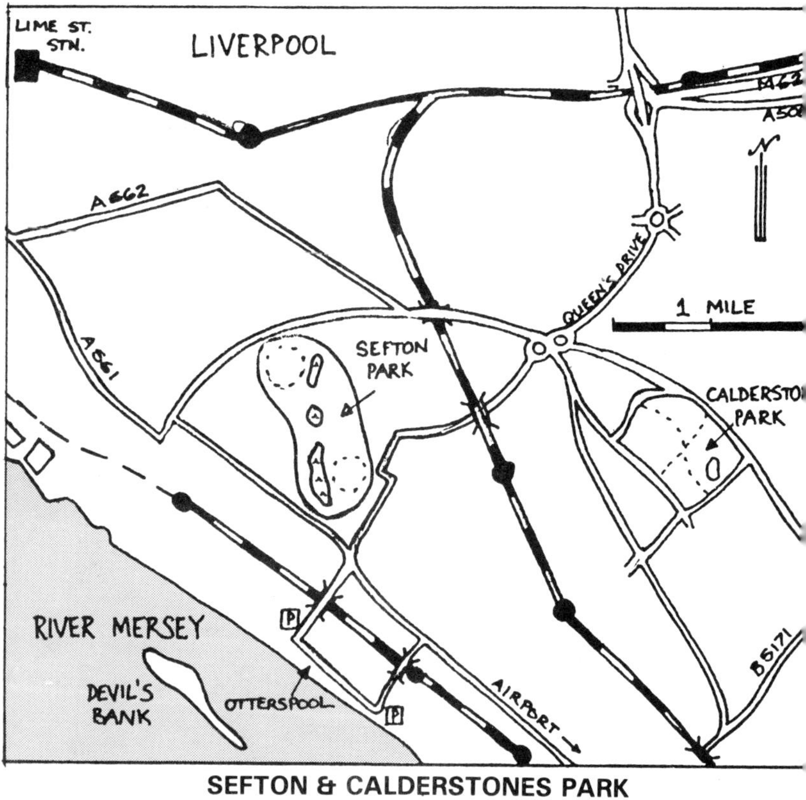

SEFTON & CALDERSTONES PARK OTTERSPOOL PROMENADE

CALDERSTONES PARK

☛ Public park, W.C. Dis. (W.C.)

LOCATION
S.E. area of Liverpool (108) SJ 407878.

ACCESS
Entrances in Menlove Ave. (A562), Yew Tree Rd., Harthill Rd., Allerton Rd.

HABITAT
City Parkland with small lake, tended gardens, mature trees.

BIRDS
An escape from the surrounding city, the park provides pleasant walks with a representative variety of commoner species. Many parkland and garden species reside, including **Mistle Thrush**, **Great Tit** and **Treecreeper. Mallards** live up to their city image as a 'park lake duck' and even breed in the gardens of some of the larger surrounding property. **Redwing** and **Fieldfare** occur in winter with the former on the grassland and along the surrounding roadway verges in substantial numbers during harder weather. **Bramblings** are attracted to the beechmast beneath the many fine beech trees in and around the park, in numbers which vary from twos and threes to flocks of several hundred, due to their irruptive nature.

GENERAL
The formal Garden displays are excellent and include Traditional, Walled and Japanese Gardens. As well as being well-planted with many large trees the grounds also hold a huge variety of Rhododendrons and Azaleas which form one of the best collections in the country.

HALE HEAD & OGLET SHORE

☛ Footpaths, minor roads.

LOCATION
North bank of River Mersey between Liverpool and Widnes (108) SJ 473814.

ACCESS
From A561 via unclassified roads from Speke, Halewood or Ditton to Hale village green. Park at Childe of Hale P.H. then follow Church Lane at the side of the pub through a gate to Mersey Way footpath and Hale Lighthouse. The path running to the right follows the shore to Oglet and the boundary of Liverpool Airport and the path to the left skirts the shoreline of Hale Head before returning to Hale church via Within Way.

Alternatively, the unmarked paths through the deciduous woodland in Hale Park (entrance 50 yards along road from the pub car park) eventually lead to a point on the coastal path half-way between Hale Lighthouse and Oglet.

Oglet: Via Dungeon Lane. At side of 'Dove & Olive' public house on Hale Road. Car park at bottom of lane.

HABITAT
Arable farmland with scattered broadleaf plantations and hedge rows, river estuary, low earth cliffs, mud and sand.

BIRDS
Farmland species include **Skylark**, **Grey Partridge** and **Corn Buntings** alongside the estuary's waders and wildfowl. **Whimbrel** are amongst the regular migrants whilst winter brings **Shoveler** to Dungeon Bank where high tides can hold concentrations of **Pintail** in rafts of well into four figures.

Autumn waders can produce an impressive list of species. **Oystercatcher**, **Turnstone**, **Ringed Plover**, **Grey Plover**, **Golden Plover**, **Curlew**, **Bar-tailed Godwit**, **Green Sandpiper**, **Common Sandpiper**, **Redshank**, **Dunlin**, **Curlew Sandpiper** and **Sanderling** all being seen in a single visit, is by no means unheard of.

Long-eared Owls used to breed in the plantations and, with luck,

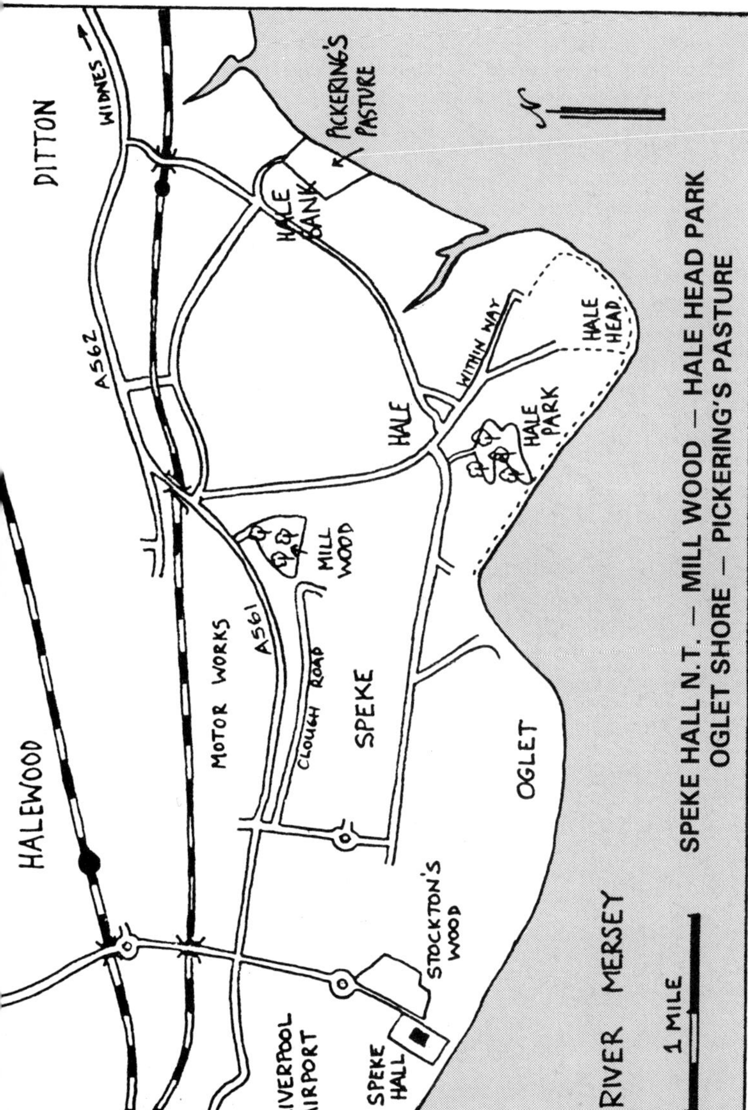

SPEKE HALL N.T. — MILL WOOD — HALE HEAD PARK
OGLET SHORE — PICKERING'S PASTURE

may do so again.

The deciduous woods of the park attract many species including resident **Great Spotted Woodpeckers**. The scarce **Mealy Redpoll** has been seen on more than one occasion accompanying a party of the **Common** variety.

The saltmarsh at Ogley is easily overlooked from the old quay below Dungeon Lane car park and is particularly good for roosting waders at high tide.

GENERAL

The Mersey Way is a well-maintained footpath which follows nearly the entire length of the north bank of the river from Hale to beyond the M6 viaduct west of Warrington. Hale itself is a picturesque village with many thatched houses and cottages; in the churchyard is the grave of John Middleton, at 9′3″ the famous local giant, the 'Childe of Hale'.

Young Barn Owls

FIDDLER'S FERRY

☛ Footpaths, Dis. (at 2).

LOCATION
North Bank of River Mersey between Widnes and Penketh (108) SJ 563868.

ACCESS
From A562.

1. From Bennett's Lane (W. end of power-station) turn left into Johnson's Lane, parking near the gate across the road. On entering the gate follow track keeping the power-station on the left hand side until reaching the canal bank. Cross to far side of canal. To the right lies an area of saltmarsh and estuary, to the left the canal-side path passes the power-station pools and leads to Fiddler's Ferry itself.
2. From Tannery Lane (E. end of power-station) turning right into Station Road across unmanned level crossing to car park. The canal-side path to the right leads to the power-station.

The power-station lagoons are private property and you should keep to the footpath.

HABITAT
Estuarine mud and sand, canal towpath, saltmarsh and reedbed.

BIRDS
The interesting combination of habitats produces a good variety of birds, principally those associated with the waterside. **Reed Buntings**, **Reed Warblers** and **Grey Wagtails** frequent the canal banks; **Mallard** and **Shoveler** inhabit the estuary, joined in winter by good quantities of waders and other wildfowl. Gulls are well-represented, the usual **Herring**, **Black-headed**, **Lesser** and **Greater Black Backed** being joined outside the breeding season by many **Common** and a few **Little Gulls** with regular winter records of **Glaucous Gull** particularly around the power-station pools. The pools also attract many of the wildfowl and waders.

The birdlife is not solely confined to wetland species, however, and a selection of farmland and hedgerow species occur. **Barn Owls** are

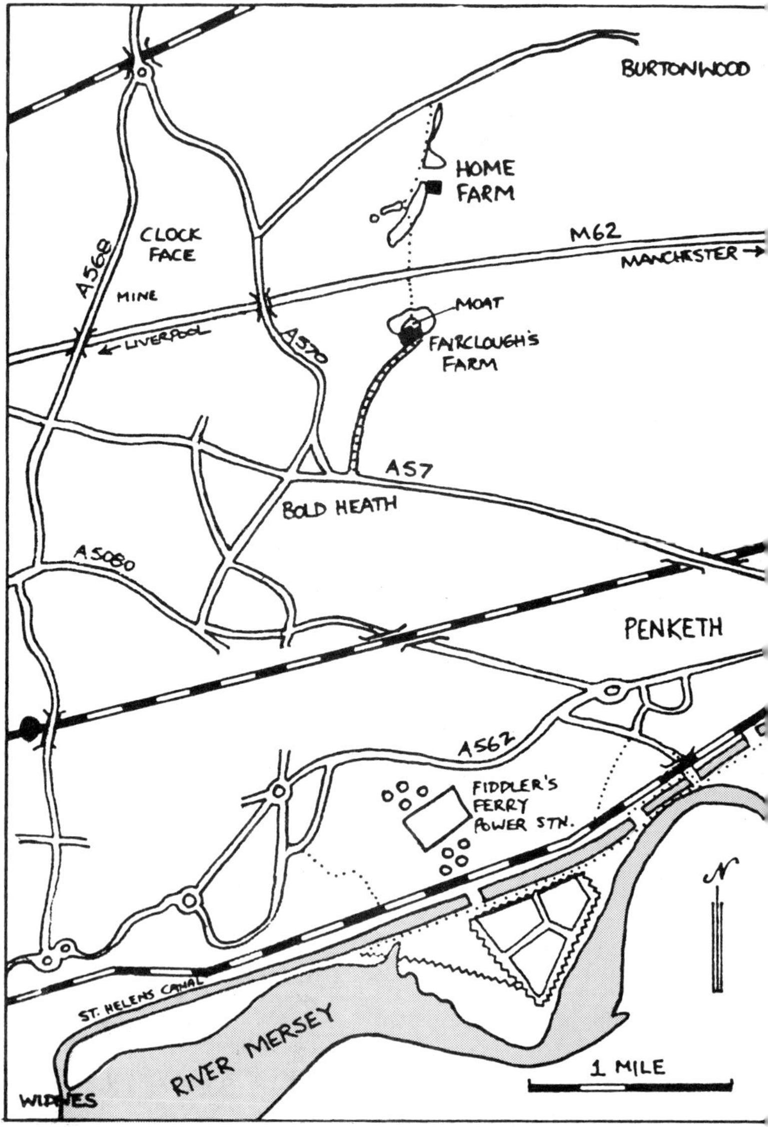

FIDDLER'S FERRY — BOLD HEATH

sometimes seen along the embankments and **Short-eared Owls** are a regular visitor from autumn to spring.

GENERAL
The Power Station pools themselves and their embankments are private land belonging to the C.E.G.B. However, visitor's permits are issued free of charge at the discretion of the Station Warden c/o Fiddlers Ferry Power Station, Widnes Road, Warrington WA5 2UT. Tel. 051-424-2020. Permit holders must park in the visitors' car park at the gatehouse and proceed on foot.

BOLD HEATH

☛ Private road used as Footpath.

LOCATION
Between Widnes and Burtonwood (108) SJ 536892.

ACCESS
From A57, park at Griffon Inn (junct. A570) cross road, then turning to the right follow road marked 'private road' on left.

HABITAT
Farmland with scattered broadleafed woodland.

BIRDS
Though the seeker after rarities is likely to be unsuccessful, this locale provides a pleasant farmland walk with the commoner hedgerow and farmland species in evidence. In summer **Corn Buntings** are to be heard constantly calling from any suitable vantage point. **Swallows** and **House Martins** breed in the farm buildings and the woodlands provide variety with **Tits** and **Warblers**, particularly in the area of the overgrown moat at Fairclough's farm. **Long-eared**, **Tawny** and **Little Owls** occur in the area.

CROXTETH HALL & COUNTRY PARK
(includes MULL WOOD L.N.R.)

☛ (Hall & Gdn. ©), Café, W.C. Dis(T), ® (L.T.N.C.) (W) Ø.

LOCATION
Liverpool, West Derby. (108) SJ 406942.

ACCESS
From A580 (East Lancashire Road) then via Lowerhouse Lane or A5048 (Queens Drive – Ring road) then via Muirhead Avenue to free car park (signposted).

HABITAT
Parkland, gardens and lawns, grazing land, deciduous woodland.

BIRDS
Lying on the edge of the main conurbation the park attracts a representative selection of field and woodland birds with over 70 species recorded to date. **Mistle Thrushes** and **Blackbirds** seek worms on the extensive lawns, the shrubbery hides **Wrens** which scold at any approach, whilst in summer **Swallows** and **Jackdaws** take advantage of the nesting opportunities afforded by the Hall itself. Near to the Hall is a small 'bird garden' where the birds are attracted to feed and the observer can watch from concealment. Perhaps proof of the old adage that anything can turn up anywhere, one of the more unusual birds to be recorded has been a **Gannet** seen flying over the park.

Many small birds like **Willow Warblers**, **Blackcaps** and **Chiffchaffs** are found all around the five hundred acres of the Park but the Mull Wood Local Nature Reserve, which is run jointly by the Park authorities, the Lancashire Trust for Nature Conservation and the Nature Conservancy Council, provides a protected area of about fifty acres for them to flourish in. A planned management scheme provides a diversity of habitat with ponds, ditches and fields harmonising with the woodland. Access to this area is generally restricted to permit holders only but a series of guided walks is arranged each spring and summer, details of which are available from the Park Rangers (see Appendix).

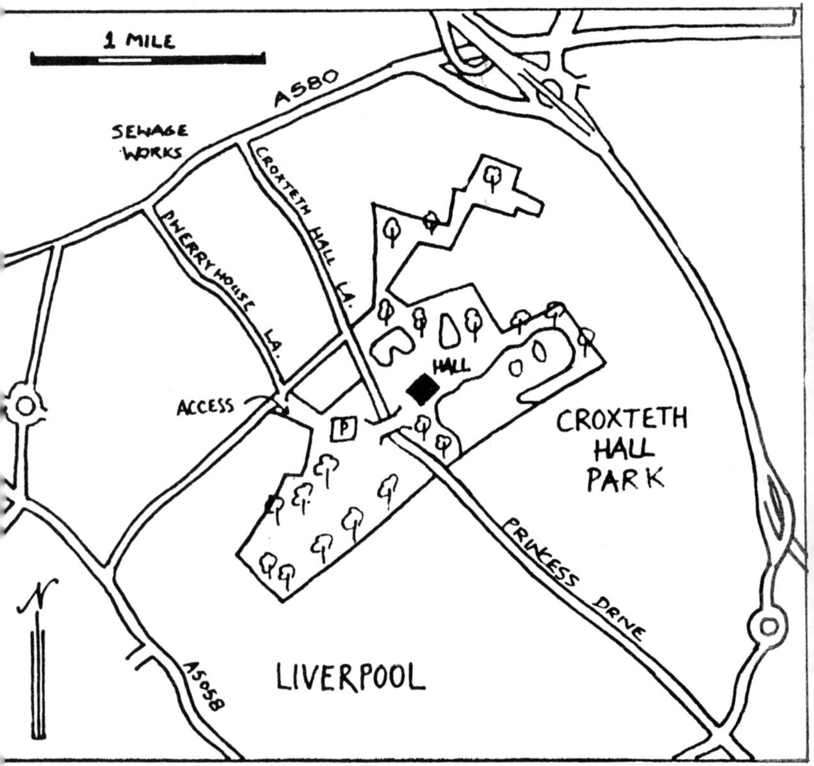

CROXTETH HALL COUNTRY PARK

GENERAL

The Parkland is open (free) daily from 9.00 a.m. to 7.00 p.m. or dusk. Admission is charged to the Hall, Walled garden and Farm. The Hall houses exhibitions of Edwardian Life (c. 1905), Carriages, Crafts and the 'Croxteth Heritage' exhibition of 1000 years of estate life, whilst the Home Farm exhibits traditional breeds and modern livestock farming as well as being an Approved Rare Breeds' Centre.

Adjacent to the Hall is the Walled Garden which contains old fruit trees, flowers, herbs and an Edwardian mushroom house.

Times of opening vary according to season.

PICKERING'S PASTURE

☛ Footpaths, ®(M.V.P.)(W), ⇌ (Ditton Junction).

LOCATION
North Bank of River Mersey between Ditton and Hale (108) SJ 488837.

ACCESS
From A561 Speke to Liverpool road via Halebank Road near junction with A562 (opposite R.S.P.C.A. Horses' Rest) to 'T' junction with Hale Road/Hale Gate Road and finally via Mersey View Road which is almost opposite.
OR – From A562 via Hale Road at the Widnes end roundabout of the Runcorn Widnes Bridge. Over railway bridge at Ditton Junction Station to junction with Mersey View Road on left.

The car park is sign posted at the end of Mersey View Road. (NOTE: The official car park closes at 4.00 p.m. each day.

HABITAT
Reclaimed tip restored to traditional meadowland, native woodland and wet meadow, bordering on estuary.

BIRDS
Commoner species like **Wren**, **Blackbird**, **Skylark**, **Corn Bunting** and **Grey Partridge** are amongst the breeding species already established and, as the restoration matures, the list will doubtless extend accordingly.

The variety of passage migrants, as well as including many passerines like **Ring Ouzel**, **Wheatear**, **Goldcrest** and **Spotted Flycatcher**, is enhanced by the proximity of the estuary. Easily observed from the path which runs along the river edge, the estuarine mud and sandbanks attract good numbers of transient waders; **Whimbrel**, **Greenshank** and **Black-tailed Godwit** figuring amongst the more numerous **Curlew**, **Redshank** and **Dunlin**. On a good day in autumn, this and nearby areas (c.f. Hale and Fiddler's Ferry) can produce as many as fourteen or fifteen wader species. It is with this in mind that a wader scrape has been excavated at one end of the site which, despite teething troubles with initially siting the hide in the wrong place, should prove worthwhile.

Winter sees the return of the wildfowl vying for attention in opposition to the **Winter Thrushes** which join the **Blackbirds**, **Song** and **Mistle Thrushes** seeking food and shelter in the fields and hedgerows, while **Finches** and **Buntings** hunt for seeds along the tideline and bankside vegetation.

GENERAL

One of the prime objects of the site, which is managed by the Mersey Valley Partnership, is to provide suitable conditions for native plants and trees, which are under pressure from modern farming practices. Meadow and hedgerow plants like Corncockle, Cornflower, Poppy, Stitchwort and Foxglove are being seeded and allowed to naturalise with this end in view.

Pied Flycatcher (m)

MILL WOOD L.N.R.

☛ Footpaths, Dis, ® (L.U.W.U.) (W).

LOCATION

South Liverpool between Speke housing estate, Ford Motor Factory and Hale village. (108) SJ 454837.

ACCESS

Speke A561 Liverpool/Widnes road. At the Western Avenue junction at Speke, turn into Western Avenue taking first left (Clough Road) then straight on to Mill Wood entrance as road bends right.

HABITAT

Oak and native woodland with brook and pond, bordering on farmland to the south and urban area to the north.

BIRDS

Many common woodland species are resident including **Blackbirds**, **Song** and **Mistle Thrushes, Great Spotted Woodpecker, Treecreeper** and **Nuthatch** as well as **Blue**, **Coal** and **Great Tits**. These resident breeders are joined by summer visitors like **Garden Warbler**, **Willow Warbler** and **Blackcap**; which also breed here, **Garden Warblers** being particularly common. Additionally in summer, visiting **Swifts**, **Swallows** and **House Martins** are to be found, capitalising on the abundant insect population the area holds, and nesting in and on nearby buildings. The adjacent farmland is home to numerous **Skylarks**; with their aerial phraseless songs often heard alongside the more jingling and repetitive notes of perch-singing **Yellowhammers**.

As the seasons change the **Warblers** and **Hirundines** are replaced by **Fieldfares** and **Redwings**, which find winter shelter and food in the wood and nearby fields while mixed parties of **Tits** patrol the trees in their "here one minute – gone the next" fashion.

GENERAL

The wood, which also holds an interesting and varied small mammal and invertebrate population, has been maintained and improved by 'Landlife' Urban Wildlife Unit (formerly the Rural Preservation Association). The pond has been enlarged and contoured to provide fishing for the local anglers.

SEAFORTH DOCK POOLS N.R.

☛ (pedestrian), W.C. Dis., Ø (vehicle), ® (L.T.N.C.) (W), ⇌ (Seaforth).

LOCATION
Seaforth (N. Liverpool) (108) SJ 324972.

ACCESS
From Crosby Road South (A565 Liverpool/Southport road) junction with Princess Way (A5036 Netherton/Seaforth road) via Mersey Dock and Harbour Company Freeport entrance, follow perimeter road keeping right, to pumping station compound near Port Radar Station. Pedestrian access only except disabled badge holders or parties by prior arrangement with reserve office – 051-920-3769.

HABITAT
Saltwater and freshwater lagoons surrounded by derelict land brought under management. Sea-wall embankment and estuary.

BIRDS
The pools, which were excavated to provide reservoirs for topping up the docks, attract many seabirds and wildfowl, the variety being in no small part due to the fact that the inner lagoon contains freshwater which provides a choice of habitat. Whilst over 180 species have been recorded the reserve is perhaps best known for its two specialities – **Little** and **Mediterranean Gulls**. The former is a winter visitor and passage migrant, several hundred being commonly present in spring. **Mediterranean Gulls** are less common but are seen practically daily between August and April, usually around high tide when they join with other gulls on the saltwater lagoon, foraging around the pumphouse inlet which is active at this time.

A substantial hide has been built overlooking the freshwater lagoon, giving excellent opportunities to observe the wildfowl with **Mallard**, **Pochard** and **Tufted Ducks** common, **Goldeneye**, **Scaup** and **Teal** regular and the occasional record like **Ruddy Duck** and **Scoter**.

Large numbers of **'Comic' Terns** occur on passage, together with numerous **Little Terns** and several wader species. The list of breeding species is limited at the moment, but as shrubs are planted and become established several species which are only recorded as

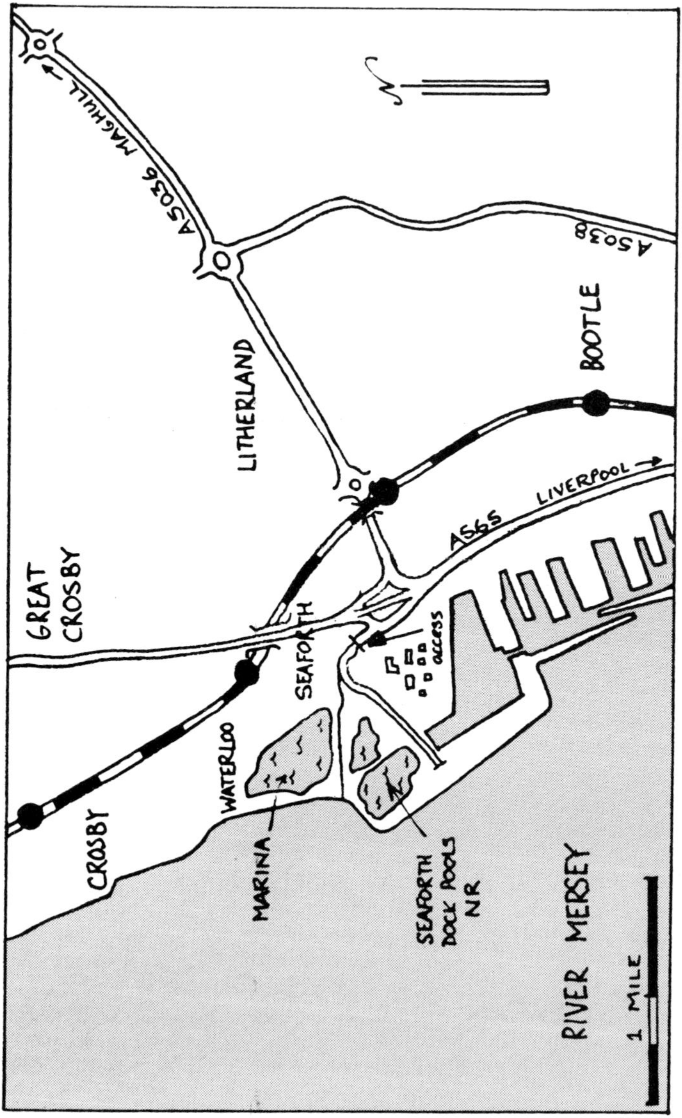

SEAFORTH DOCK POOLS

Pochard

visitors at present will add to it.

GENERAL

The reserve, which is owned by the Mersey Docks and Harbour Company, is managed and wardened by the Lancashire Trust for Nature Conservation.

OTTERSPOOL PROMENADE

☛ Public park and paths, Dis., ⇄ (Otterspool).

LOCATION
S.W. Liverpool (108) SJ 375860.

ACCESS
From Aigburth Road (A561 Liverpool/Widnes road) via (from south) Jerico Lane or (from north) Mersey Road, to car parks on Otterspool Drive.

HABITAT
Parkland and shrubbery alongside river estuary with mud and sandbanks.

BIRDS
The parkland provides a home for many of the commoner residents like **Wren, Blackbird** and **Blue-tit. Pied Wagtails** are commonly seen picking their way along the pathways and in winter the shrubbery and grassland attracts **Fieldfares** and **Redwings**, often in large numbers, to join the ever-present **Song** and **Mistle Thrushes**. Tired migrants often seek shelter in the shrubs which are accordingly worth a second look in both spring and autumn. The broad expanse of the Mersey is never far from mind or view and draws the observer like a magnet particularly in autumn and winter. Autumn brings concentrations of **Great-crested Grebes**, often more than a hundred, moving with the tide as it surrounds and covers Devil's Bank, finally displacing the many **Cormorants** which find it a convenient spot to spread and dry their wings. The **Grebes** are present in the winter, though in reduced numbers, when the wildfowl are at their peak with quantities of **Mallard** and **Teal** and the thousands of **Pintail** which make the estuary internationally significant as a winter refuge. Perhaps the optimum place to observe birds closely is the sewage outfall at the north end of the promenade – best at half-flood tide, as a portion of the shoreline is still exposed and many different species of waders can then be seen alongside the wildfowl. For the beginner it is perhaps a useful spot to come to close quarters with the 'five' – **Greater Black-backed**, **Lesser Black-backed**, **Herring**, **Common** and **Black-**

Pintail

headed Gulls, in all plumage phases and really get them sorted out from one another.

GENERAL

The site is reclaimed from a rubbish tip and lies immediately next to the International Garden Centre site which also is on land reclaimed from the same tip when it finally ceased to operate.

SPEKE HALL

☛ 0 ©, Café, W.C. Dis. (W) N.T.

LOCATION
South Liverpool near Airport (108) SJ 423832.

ACCESS
From A561 Speke Rd. (Liverpool/Widnes road) via Speke Hall Avenue (end of airport). Follow road straight on when main road bends sharp left.

HABITAT
Mature deciduous woodland, ponds, lawns and gardens bordered by embankment and arable fields.

BIRDS
The Hall and its grounds provide an edge of the city oasis of 68 acres which includes 30 acres of prime habitat in the form of Stockton Wood where **Sparrowhawks** and **Jays** are notable amongst the regular breeding woodland birds. The lawns and managed gardens provide diversity, and **Pheasants** still breed amongst the rhododendrons which were introduced to provide cover for gamebirds, although modern management is seeking to reduce these foreigners which interfere with the growth of many of our native species. The embankment to the south of the Hall separates it from the Airport taxiway which cuts off the former access to the foreshore which used to provide additional interest in the form of waders and wildfowl (see p.44 Hale Head/Oglet), but compensates by providing a breeding habitat for **Meadow Pipits** and **Corn Buntings** and a hunting ground for **Kestrels**.

GENERAL
Closed between the end of December and April 1st and on Mondays each week (except Bank Holiday Mondays), the house and grounds are open as follows:–
April 1st to October 31st.
1.00 p.m. – 5.30 p.m. Tuesday to Saturday.
12.00 noon – 6.00 p.m. Sundays and Bank Holiday Mondays.
November 1st to December 20th
1.00 p.m. – 5.00 p.m. Saturdays and Sundays only.

Grey Heron

WIGAN FLASHES

☛ Towpath, Footpaths ⇄ (Wigan).

LOCATION
South of Wigan (108) SD 578045.

ACCESS
From A49 Warrington/Wigan road taking B5238 Poolstock road at roundabout.

From A573 Manchester/Wigan road via A577 (town centre) then Poolstock road.

Park near Poolstock Church and cross main road to canal towpath entrance immediately opposite, turning right along either bank of canal to the flashes.

HABITAT
A dozen or more fresh-water colliery flashes of varying size and depth, Scotsman's and Pearson's being the first and second largest respectively, surrounded by shale embankments with some low vegetation, willow and birch.

BIRDS
Waterfowl are in abundance all year round with breeding **Mute Swans**, **Tufted Duck**, **Mallard** and **Kingfisher**. **Grey Herons** are common and the scarce melanic form of this species has been recorded here. **Blackcap** and **Golden Warblers** frequent the scrub and shrubland and other breeding passerines number **Reed Bunting** and **Grey Wagtails** amongst them.

The local sailing club shares Scotsman's Flash in reasonable harmony with the birds, certainly not bothering the gulls which are ever-present, all of which provides a background to the frequent interesting records of visitors like **Black-throated Diver**, **Slavonian Grebe**, **Grey Phalarope** and **White-winged Black Tern**.

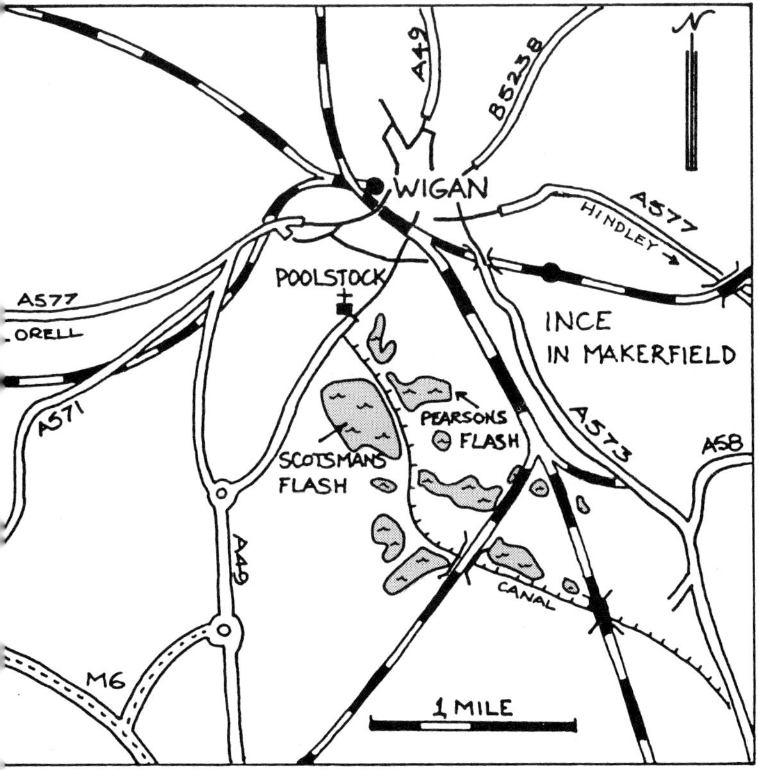

WIGAN FLASHES

SIDINGS LANE, RAINFORD

☛ W.C.* Ⓡ (St. Helens M.B.C.) (W).

LOCATION
S.W. Lancashire between Ormskirk and St. Helens (108) SD 467022.

ACCESS
A570(T) Ormskirk/St. Helens road. From M58 Junction 3 or from A580(T) Rainford turn-off. Sidings Lane runs west immediately alongside the north side of the railway-bridge crossing the A570(T) at Rainford.

HABITAT
Birch and alder woodland bordered by railway embankment and arable land.

BIRDS
In spring, the surrounding farmland often holds the first of the early migrant **Wheatears** as the resident male **Woodcocks** perform their bat-winged 'roding' flight, uttering their gutteral calls which can be heard from up to four hundred yards away. **Spotted Flycatchers** and **Cuckoos** are also among the summer visitors, the latter principally parasitising the local, ground-nesting **Meadow Pipits**.

Resident woodland species are well-rêpresented by **Jay**, **Woodpigeon**, **Wren**, **Great** and **Blue Tits** joined in summer by **Willow Warblers** and **Chiff-Chaffs**. The path to the left of the central lane or track winds through the wood and parallels the railway. The embankment scrub and shrub provides a mini-habitat for several species to breed and feed in, including an opportunity for **Long-tailed Tits** to find secure sites in which to build their elastic, moss and gossamer nests.

GENERAL
The area is managed by St. Helens Council Dept. of Leisure Services Ranger Service and the site provides a car park and picnic area. See also Sankey Valley Park p.

NOTE:
* The Ranger Station and toilets are normally closed except at weekends.
** Public access is restricted to the area between the lane and the railway.

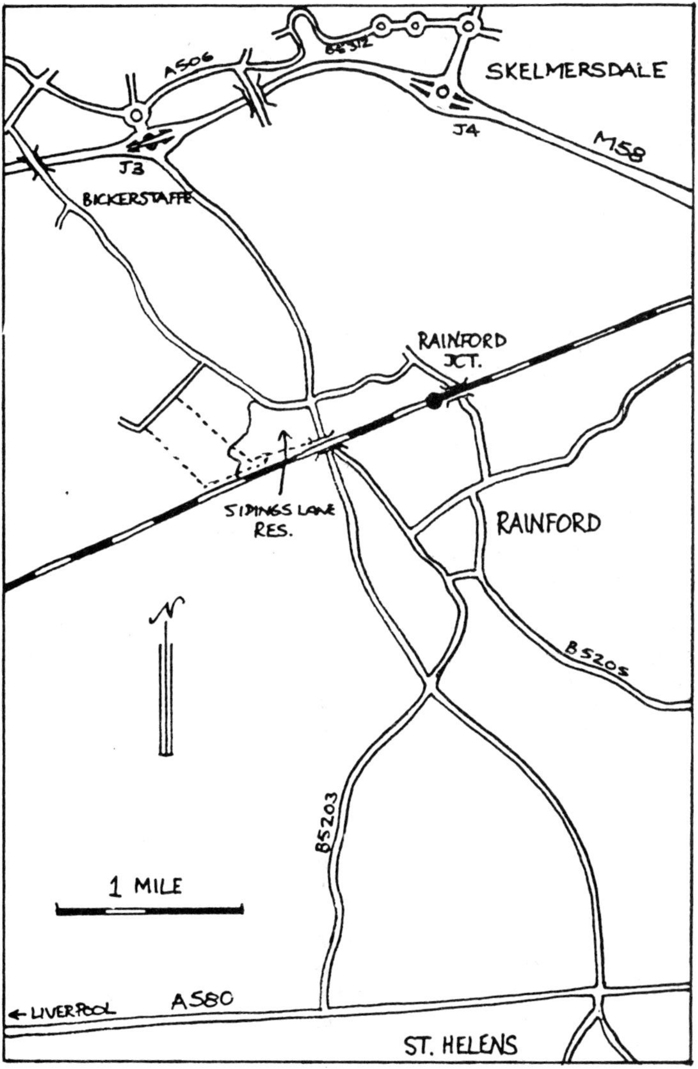

RAINFORD SIDINGS LANE

SEFTON PARK

☛ Public Park, Café, W.C., Dis.(T).

LOCATION
South Central Liverpool, bordered by Ullet Road, Aighburth Road, Aighburth Vale and Greenbank Road. (108) SJ 380879.

ACCESS
By way of the numerous minor roads leading from the above.

HABITAT
City parkland with connected lakes, tended gardens, shrubbery, playing fields and mature trees.

BIRDS
The preponderance of native chestnut, oak, beech and ash provide a home for **Nuthatches** and **Treecreepers**. Here **Great Spotted Woodpeckers** demonstrate their ability, given suitable circumstances, to penetrate deep into the urban zone and live quite happily surrounded by man. The numerous pathways, including a numbered nature trail, meander through the park but one is never far away from the waterway and system of lakes which form a principal feature and attract visiting wildfowl like **Pochard**, **Tufted Duck** and **Canada Geese** to join the ubiquitous **Mallards** with their seemingly insatiable appetite for donations of bread.

Winter brings the visiting **Redwings** to join **Mistle Thrushes** fossicking for food in the grass and leaf-mould and the abundance of sticklebacks in the lakes has attracted a **Kingfisher** on more than one occasion.

An early morning mid-week visit, when human visitors are few, often pays dividends, particularly with the more easily disturbed visiting wildfowl.

GENERAL
Sefton is one of the three large municipal parks created during the mid 1800s for the amenity of the working-classes. It was once part of the Royal Deer Park of Toxteth which surrounded the main hunting lodge, Park Lodge, which still stands at the end of Lodge Lane.

Greenbank Park which adjoins the S.E. edge of Sefton Park also contains a small lake and mature trees and many of the Sefton birds extend into this habitat.

SANKEY VALLEY PARK

☛ Footpaths, W.C., Dis(T), Ⓡ (St Helens M.B.C.) (W), ⇌ (Earlstown).

LOCATION
South Lancashire, 10 miles of linear paths extending south-eastwards from Billinge Beacon near St. Helens to the Cheshire border south of Newton-le-Willows (108) SD 535966.

ACCESS
There are numerous Pedestrian accesses along the full length of the Park.
Parking is available at any of the following access points:–
(i) Carr Mill Dam, junction A580(T) Liverpool/Manchester road and A571 Billinge road.
(ii) Park Visitor Centre (A58 St. Helens/Ashton-in-Makerfield road rear of "Ship Inn", Blackbrook).
(iii) Havannah Lane (off A572 St. Helens/Newton-le-Willows road at the "Engine Inn").
(iv) Warf Road, Earlestown, Newton-le-Willows (off A572).
(v) Vulcan Village, Newton-le-Willows (off Wargrave Road, signposted Red Brow Wood car park.

HABITAT
Descending from Billinge Beacon the park passes alongside arable fields and skirts Car Mill Dam Reservoir before continuing through the typical riverine valley-bottom. Ash, willow and birch with patches of other, principally deciduous, woodland occur at intervals along its length. The Sankey Brook is accompanied along much of its length by the St. Helens Canal which conjointly provides a wide spectrum of water and bankside habitat.

BIRDS
The extent of the Park, passing as it does, through industrial, suburban and rural areas, supports 129 species which have been recorded over the past five years.

All five **Owls** occur with **Barn**, **Tawny** and **Little Owl** breeding. The smaller woodland birds are well-represented; breeding **Blackcap**, **Willow** and **Wood Warblers** augmented by a plentiful supply of

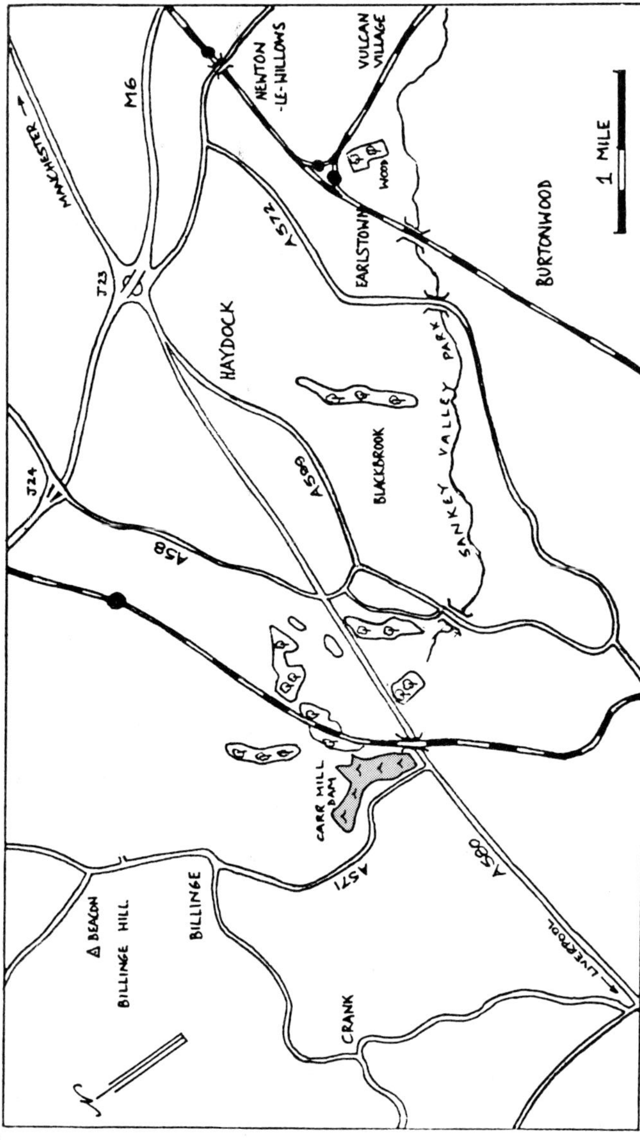

SANKEY VALLEY PARK

resident **Goldcrests** and visiting **Siskins**. In some areas the **Willow Tit** is the commonest of the **Titmice** with its harsh 'chyearing' call standing out from the less grating notes of the **Long-tailed**, **Coal**, **Blue** or **Great Tits**, all of which are residents.

The waterside attracts **Grey Wagtails**, **Water Rails** and even, on one occasion a **Dipper**. As many as five pairs of **Great Crested Grebe** have bred on Carr Mill Dam, despite the presence of power boats. Elsewhere, **Kingfishers** are present throughout the year and are suspected of having bred. Winter sees many additional gulls and wildfowl visiting the Dam area including scarcer species like **Goldeneye**, **Goosander** and **Gadwall**.

Among the several rarer birds to have been seen in the Park are **Great Northern Diver**, **Black Tern**, **Alpine Swift**, and **Yellow-browed Warbler**.

GENERAL

The Park is managed by St. Helens Council Dept. of Leisure Services Ranger Service (See also Sidings Lane p.64) and the site provides a visitor centre and picnic areas.

Access is available for wheelchairs – please contact the visitor centre for information (see Glossary).

Goosander

PENNINGTON FLASH COUNTRY PARK

☛ Footpaths, W.C., Public Hides, Dis. (T), Ⓡ (Wigan M.B.C.) (W).

LOCATION
Lancashire S.E. of Wigan (109) SJ 647986.

ACCESS
Signposted from A580 East Lancashire Road junction with A572 Leigh road. A turning opposite Railway Inn leads to Information Centre and Car Parks.

HABITAT
Over 1000 acres consisting principally of a freshwater colliery flash surrounded by grassland, scrub and managed parkland, trees and shrubs.

BIRDS
Planned management has resulted in more and more species being recorded in an area originally best known for water-fowl. The list now extends to more than 200 species. Inevitably, the large expanse of water still dominates and species like **Little Grebe**, **Mute Swan**, **Shoveler** and **Tufted Duck** breed. The resident **Great Crested Grebes** are often joined by visiting birds which result in regular counts of over a hundred individuals. **Scaup** and **Common Scoter** are passage migrants and **Whooper Swans**, **Wigeon**, and **Pintail** occur in winter.

The land surrounding the flash is rich in both resident and visiting bird-life, the more notable regulars including **Lesser-Whitethroat**, **Kingfisher** and **Jack Snipe. Iceland** and **Glaucous Gull** are recorded annually and a liberal scattering of rarities, which include species like **Goshawk**, **Purple Heron** and **Red-Crested Pochard**, makes Pennington one of the best inland sites in the region.

GENERAL
The park is managed by Wigan Metropolitan Borough Council and offers additional facilities for watersports and fishing. An on-site Information Centre provides information on the history of the area as well as its interesting plant, insect and animal life.

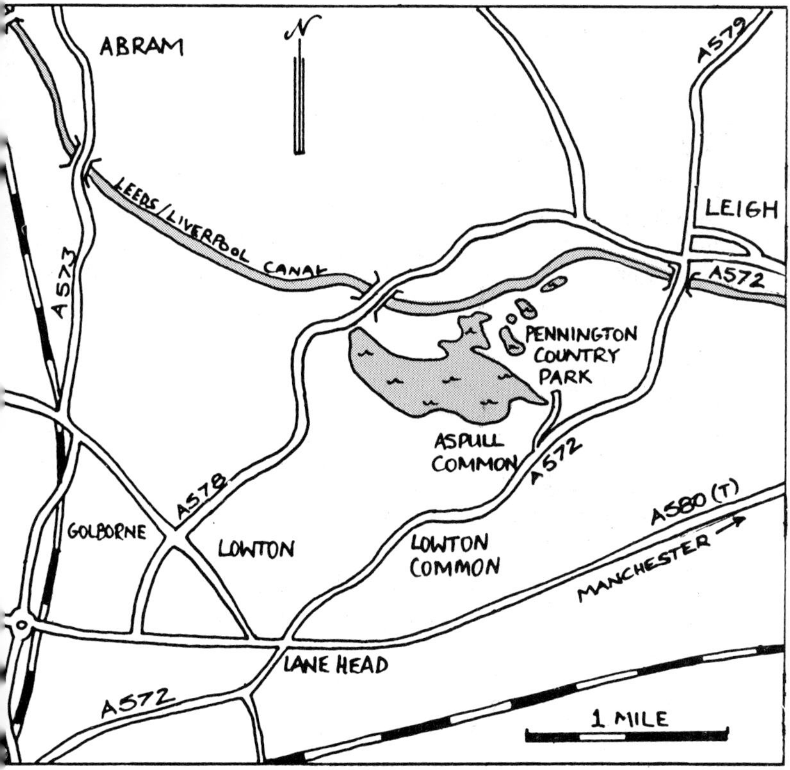

PENNINGTON FLASH

Grey Phalarope

SECTION 3
MERSEY to DEE

WEAVER BENDS & FRODSHAM MARSH

☛ Footpaths & tracks, ⇌ (Frodsham).

LOCATION

South of River Mersey between Runcorn and Helsby. (117) SJ 520786.

ACCESS

1. From A56(T) Warrington/Chester Road via St. Hilda's Drive (E. side of town on brow of hill near church with spire). The road bends gently right becoming Hayes Crescent, then sharply left. Take the next right and then right again at the 'T' junction. Finally, take the minor road on the left which crosses the motorway, continuing along the unmetalled stretch to park near the gate across the road. On entering the gate turn right to follow the footpath around the sludge-bed or go straight on to reach the embankment path.
2. From A56(T) Warrington/Chester Road via Marsh Lane (W. side of town near railway bridge). Crossing the motorway follow unmetalled road to right; after road rises parking area is on the right. Follow footpath along ridge to embankment or descend to path connecting with above embankment.

No suitable access for coaches beyond motorway bridges.

HABITAT

Estuary and bankside, arable land, hedgerow and embanked chemical sludgebeds.

BIRDS

Regular visitors and breeding birds include **Sedge Warbler**, **Reed Warbler**, **Little Ringed Plover**, **Ringed Plover**, **Reed Bunting** and **Grey Partridge**. Passage migrants in great variety in both spring and autumn are attracted to the area and the chemical sludgebeds can hold large numbers of waders which often include rarities such as **Bairds** and **Pectoral Sandpipers**, **Red-necked** and **Grey Phalarope**, **Little** and **Temmincks Stint** all recorded from time to time and **Wilson's Phalarope** on at least one occasion.

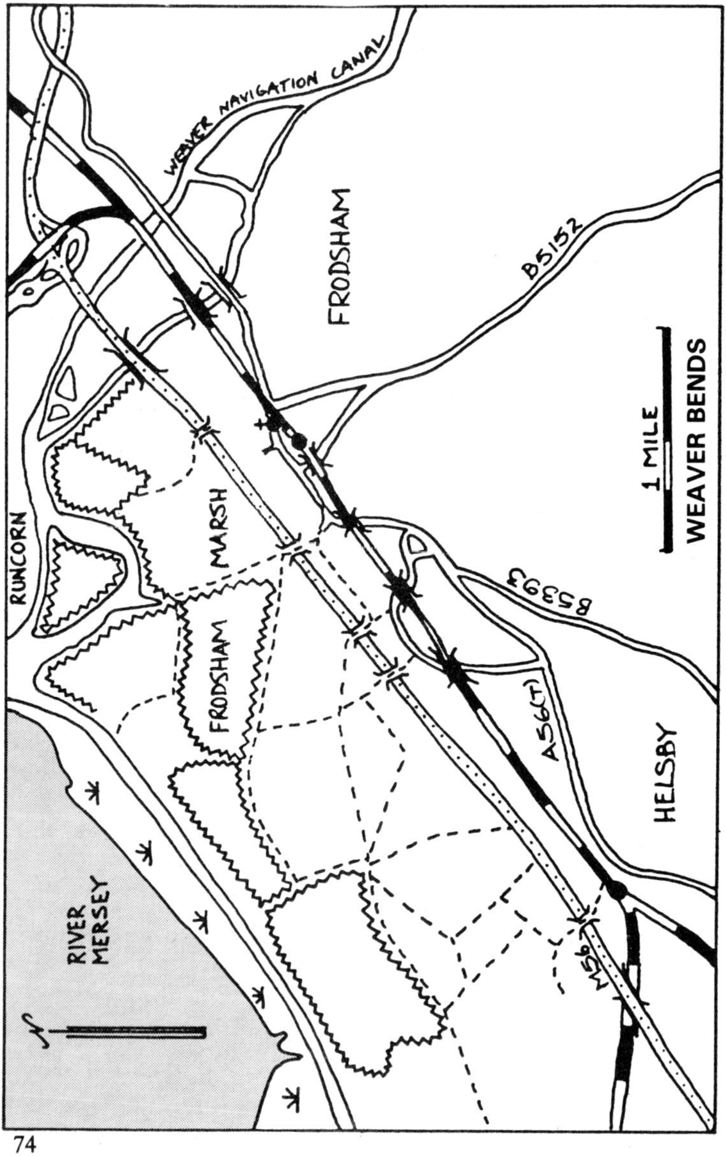
WEAVER NAVIGATION CANAL
FRODSHAM
B5152
1 MILE
WEAVER BENDS
RUNCORN
FRODSHAM
MARSH
B5393
A56(T)
HELSBY
M56
RIVER MERSEY

Reeve (f)

Winter wildfowl in good variety include **Mallard, Teal, Shoveler, Tufted Duck**, with **Smew** and **Long-tailed Duck** as an occasional attraction. **Short-eared Owls** are in almost constant attendance during winter months, showing up well in the oblique light of a clear afternoon, as they hunt over the embankments and surrounding fields. Personal records include 10 duck, 2 goose and 2 swan species, all in one 400 metre stretch during one particular period of harsh weather.

GENERAL

Many of the tracks crossing the marsh and farmland are public footpaths but some are closed to vehicles.

Warning: The sludgebeds are private property, also what may look to be a solid surface is in fact deep, soft, mud and should not be walked upon.

DELAMERE FOREST

☛ Footpaths, tracks, W.C., DIS (W.C.), ⥂ (Delamere).

LOCATION
N.W. Cheshire (117)SJ 555722.

ACCESS
A56(T) Warrington/Chester Road via B5152 (Frodsham), taking care not to miss the point where the road turns right after a dip (signposted Delamere)
OR A556(T) Northwich/Chester Road via B5152 opposite Vale Royal Abbey Arms.
(i) Car Park at lakeside (Hatchmere).
(ii) Minor Road at side of Forest Café & Shop – Car Park ¾ mile on left.
(iii) Delamere Station (disused) Car Park & Information Centre.

HABITAT
Pineforest with some deciduous belts (mainly peripheral), plantations, clearings and mere.

BIRDS
Hatchmere is well known for its breeding pairs of **Great Crested Grebes** which re-appear during the same three or four day period towards the end of February each year. **Pochard**, **Tufted Duck**, **Reed Warbler** and **Little Grebe** are all to be found on or around the Mere whilst the Forest numbers **Green**, **Great-spotted** and **Lesser-spotted Woodpeckers**, **Jay** and **Goldcrest** among its resident attractions. **Sand Martins** have bred in the area and **Crossbills** are regularly recorded although are not thought to have bred.

Many footpaths cross the forest and waymarked trails run from the Station Information Centre. The Car Park at (ii) is a regular winter spot for many of the woodland species especially **Great-spotted Woodpeckers**, **Blue Tits**, **Coal Tits**, **Great Tits** and **Nuthatches** – all of which are remarkably bold, at times, in coming to be fed by visitors.

GENERAL
Part of the 'Sandstone Trail', a 30 mile long waymarked footpath

stretching from Frodsham to the Shropshire Border, runs through the forest and can be joined at the Car Park at (ii) above.

The forest floor can be very dry and care should be taken not to start fires!

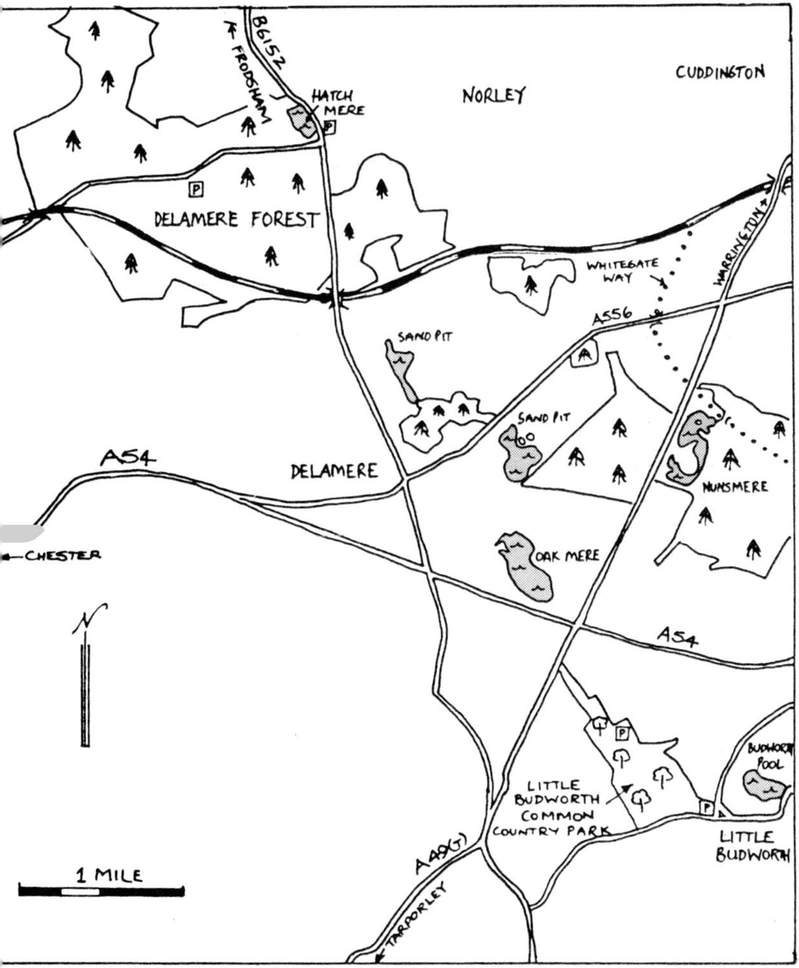

DELAMERE FOREST — NUNSMERE — BUDWORTH POOL — BUDWORTH COMMON C.P.

STANLOW N.R.

®, Ø, (P), (M.N.A.).

LOCATION
South Bank of the River Mersey extending from Stanlow Point Gowey Estuary to Mount Manisty Pool Hall Gutter. (117) SJ 429769.

ACCESS
By party permit only: obtainable from Merseyside Naturalists' Association, (see appendix). Experienced observers may arrange to visit accompanying wildfowl counting team. Entrance to the reserve is at present gained through the Shell Oil Company complex at Stanlow, crossing the Manchester Ship Canal on the Company's ferry at Stanlow Point whilst this remains in service. There is NO public access to the reserve.

HABITAT
Tidal saltmarsh.

BIRDS
Established by the M.N.A. in 1970 the reserve is designated as an S.S.S.I. (Grade 1.) due to its prime importance to wildfowl. The area boasts the country's largest estuarine counts of **Pintail** and **Teal**, and the second largest national total of **Shelduck**, as well as notable numbers of waders like **Redshank** and **Golden Plover** on occasion.

Permit holders may walk from the ferry along the top of the canal embankment where the vegetation and shrubbery provide nesting haunts for **Reed Buntings**, **Sedge Warblers** and **Pied Wagtails**. One or two larger trees, particularly around Mount Manisty, harbour the **Carrion Crows** and **Magpies** which feed out on the marsh. The Mount, named after one of the Ship Canal's surveyors, was formed from the excavated material from the canal and now it has naturalised with the landscape, makes an excellent vantage point from which to observe birds. **Grey Herons**, often numbering in double figures, roost on the outer side of the canal embankment just past the northwest end of the reserve.

High tides cover the marsh and in winter the wildfowl form rafts in and around Manisty Bay until the waters recede to again reveal their feeding grounds.

GENERAL

Shooting in the area between the eastern end of the reserve and the Weaver sluices, comprising of Ince Banks and Frodsham Score, is controlled by the local wildfowling association.

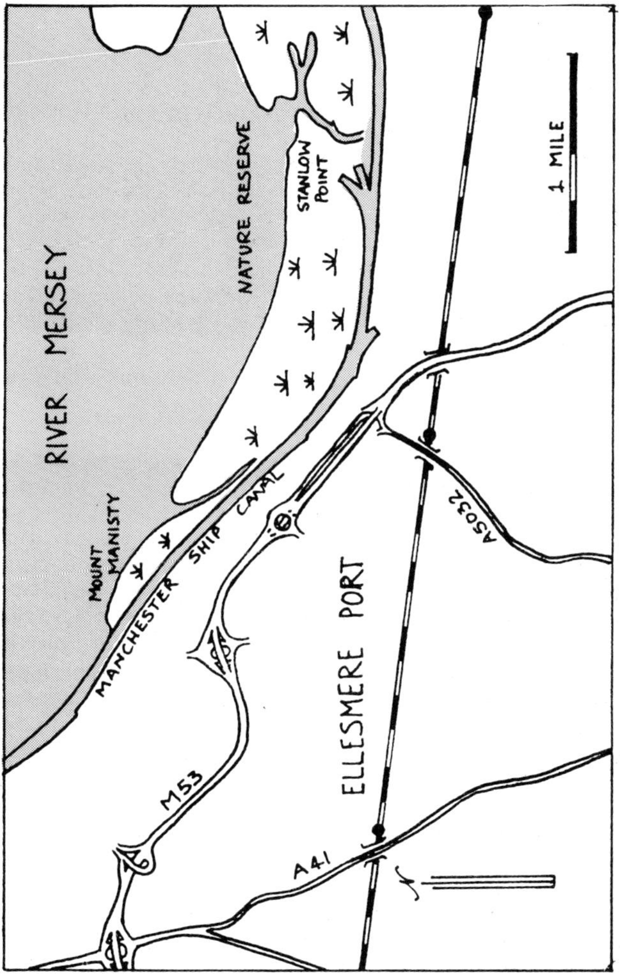

STANLOW NATURE RESERVE

WOOLSTON EYES N.R.

® (W.E.C.G.), (P), (W), ⇌ (Warrington).

LOCATION
2 miles S.E. of Warrington, Cheshire. (109) SJ 660889.

ACCESS
From M6 motorway Exit 21 take A57 for Warrington then via Weir Lane which runs south from the A57 close to the motorway junction.

HABITAT
800+ acres of wetland between the Ship Canal and the Mersey comprising of 4 large deposit beds surrounded by earth embankments. The beds are owned by the Manchester Ship Canal Co. and are used, in rotation, for disposal of dredgings from the canal which ensures a diverse habitat slowly cycling between mud and scrub.

BIRDS
Although primarily of note for winter wildfowl, including **Teal** in four figure numbers and good numbers of **Shoveler**, **Mallard** and **Pochard**; the area has attracted around 30 species of waders: highlighted by records of **Buff-breasted** and **White-rumped Sandpipers**, and **Temminck's Stint**. The peak numbers of waders occur principally on spring and autumn passage.

Summertime, whilst not providing the quantities of birds which earned the reserve its S.S.I. rating, is by no means quiet: **Shoveller**, **Mallard**, **Tufted**, **Gadwall** and **Ruddy Duck** breed here as do appreciable numbers of both resident and visiting passerines like **Sedge**, **Reed**, **Grasshopper** (10-20 pairs), **Willow** and **Blackcap Warblers**. **Lesser Whitethroat** are regularly recorded, but are not known to have bred.

Besides the rarer waders mentioned earlier, **Honey** and **Rough-legged Buzzard**, **Hobby**, **Marsh Harrier**, **Ring-necked** and **Ferruginous Duck**, **Spotted Crake** and **Laughing Gull** are among the many unusual sightings which have been recorded.

GENERAL
A public footpath runs through the reserve but access to the hides

and reserve proper is by permit only. Details may be obtained by sending an s.a.e. to Mr B. Ankers, Chairman, Woolston Eyes Conservation Group, 9 Lynton Gardens, Appleton, nr. Warrington, Cheshire.

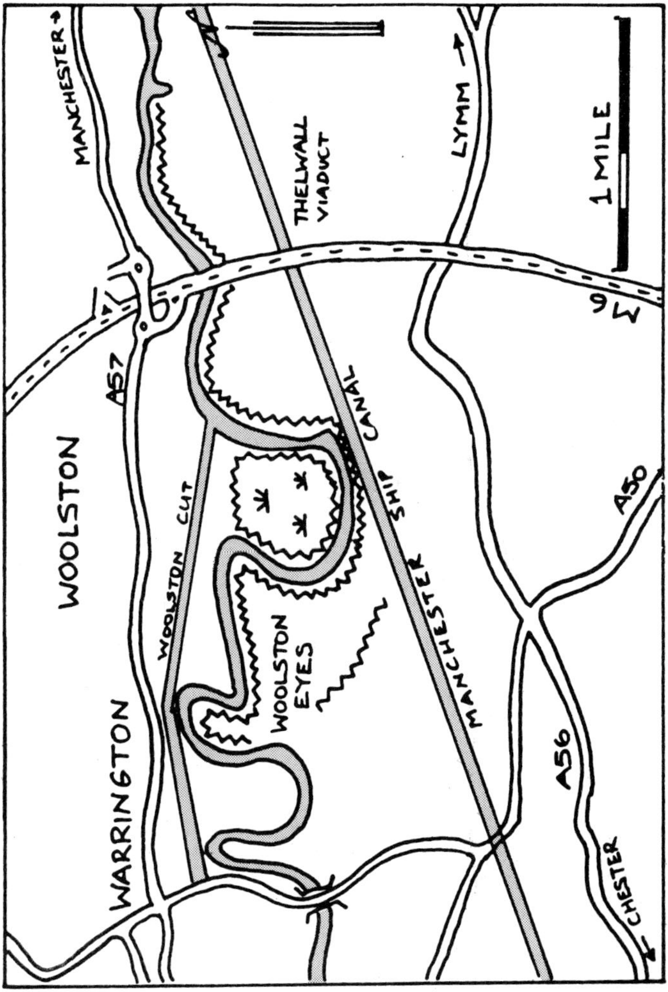

WOOLSTON EYES

NORTON PRIORY

☛, ©, P (free), Museum W.C., Dis. (W.C.), ⇌ (Runcorn).

LOCATION
Near Runcorn, Cheshire (108) SJ 548831.

ACCESS
From M56 Junction 11 turn for Warrington and follow signposts. From Widnes/Runcorn Bridge via A558, Runcorn Expressway, follow "Other Runcorn Traffic" signs to Astmore East junction, bearing left and then first right (Warrington Road), Priory is signposted on right.

HABITAT
16 acres of gardens and parkland with small stream, surrounded by mixed woodland and playing fields.

BIRDS
This site provides a varied selection with yearly total of around 70 species seen. The shrub cover is attractive as a winter roost to many birds particularly the hundreds of **Redwings** which feed on the adjoining playing fields. The shrubbery is doubtless also the factor which encourages the odd **Blackcap** to remain all year round with the more regular residents like **Bullfinch**, **Nuthatch** and **Goldcrest**. Summer brings many visitors such as **Whitethroats**, **Yellow Wagtails** and regular sightings of **Sand Martins** which have a colony on the nearby canal bank.

Besides the birds, the area has a good mammal population including Red Fox, Weasel, and Water Vole. Great Crested Newts breed in the pond and at least two species of bats are present.

GENERAL
The site of a medieval priory dissolved in 1536 by Henry VIII and then successively a Tudor mansion and Georgian country house, Norton has now been restored and developed as a Museum by the Norton Priory Museum Trust. The Priory is open daily throughout the year from 12 noon (closing times vary). The admission charge includes entrance to the museum. Parties wishing to visit outside normal hours may do so by prior arrangement.

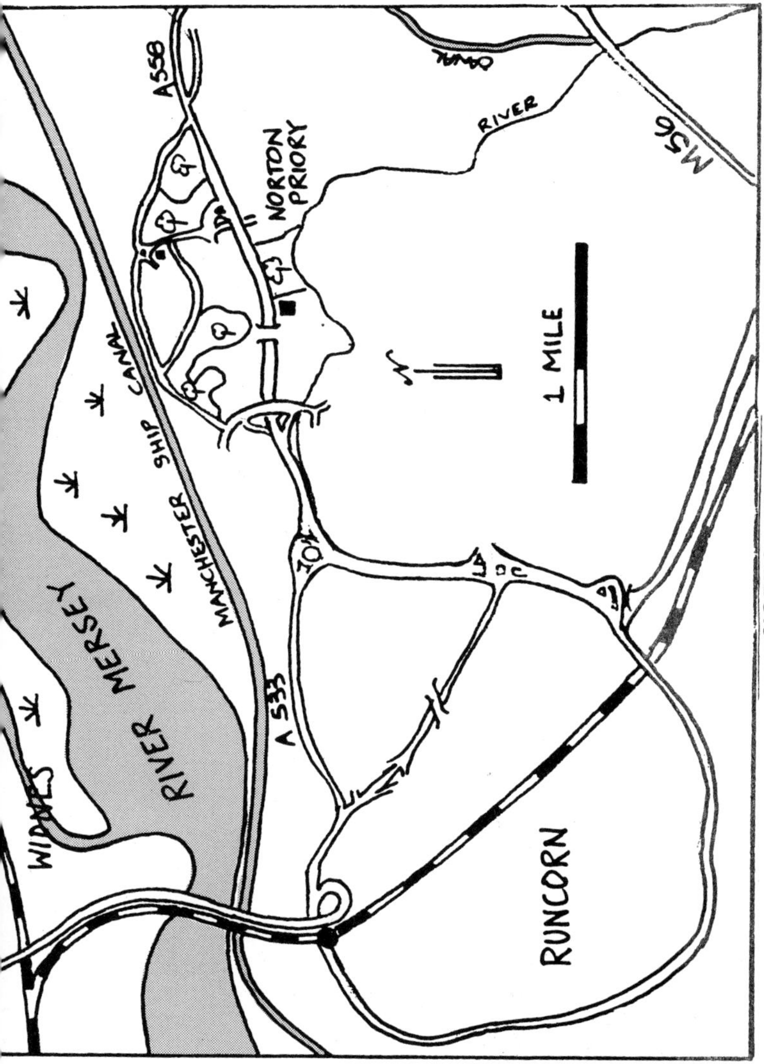

NORTON PRIORY

RANGE POOLS/SHOTWICK LAKE and SHOTTON STEELWORKS N.R.

☛ Footpaths [incl. ® (B.S.C.) (P)].

LOCATION
S.W. Wirral (117) SJ 311722.

ACCESS
From A550(T) Follow signs for Deeside Industrial Estate via perimeter road straight on for one mile, passing weighbridge. Shotwick Lake is to the right immediately before a low bridge. For range pools continue under low bridge bearing left and then sharp right over railway. For the Steelworks Reserve keep left under the low bridge and follow the road round to the right and through the works' security barrier.

HABITAT
Freshwater reservoir, scrub, freshwater pools and embankments.

Common Tern

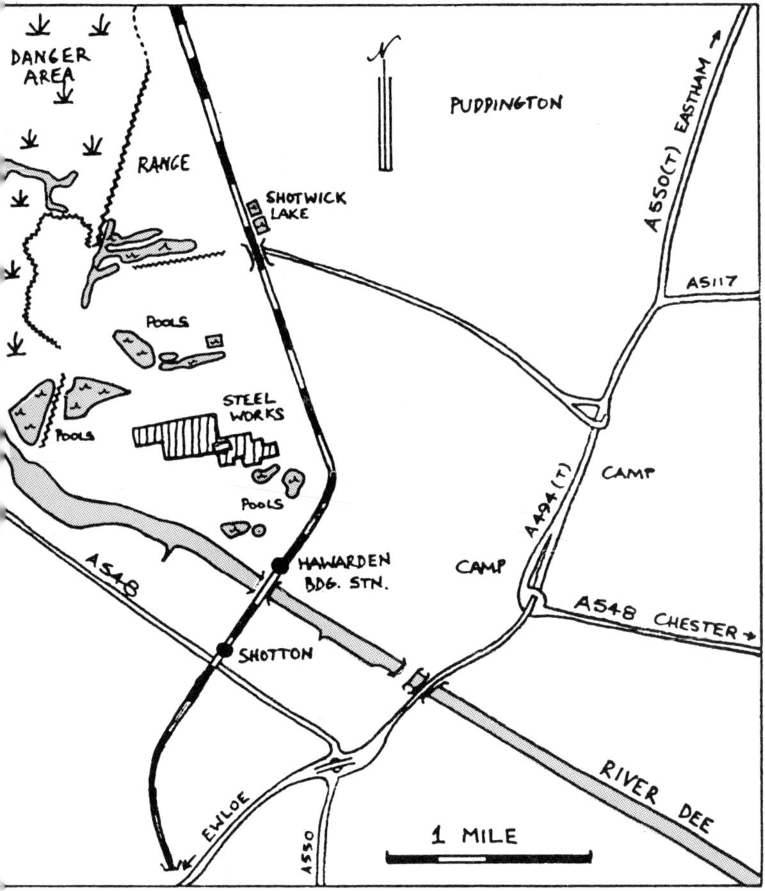

RANGE POOLS – SHOTWICK LAKE

BIRDS

Despite being used for watersports, Shotwick Lake can hold good quantities of wildfowl like **Tufted Duck**, **Mallard** and **Teal** with the lake's fringes attracting various Waders. Severe weather can concentrate species in the area; **Pink-footed Geese**, **Bewick's**, **Whooper** and **Mute Swans** visit the adjacent fields, all being present

Bewick Swan

simultaneously on at least one occasion. In summertime **Yellow Wagtails** are abundant joining the resident **Pied** and **Greys**.

On crossing the railway line, the area to the right is maintained as a reserve by The Dee Wildfowlers' Club and the range pools attract a number of interesting visitors besides the commoner species; records include **Black-necked** and **Red-necked Grebe**, **Spotted Crake** and **Crane**. A breeding colony of **Black-headed gulls** occupies one of the larger pools. Added interest may be found in the quantity of orchids, principally the striking Pyramidal variety, which may be found on the site.

The establishment of the Steelworks Reserve, which is managed by the Merseyside Ringing Group, has allowed for the building of

Yellow Wagtail (m)

artificial platforms and islands in the pools which has resulted in the formation of a substantial and thriving colony of **Common Terns**. Smaller pools, scrub and reedbeds provide a varied and secure habitat which induces many residents and summer visitors to breed.

GENERAL

The rifle range is enclosed between two embankments erected in turn as land was drained and re-claimed from the estuary. The outer embankment, which overlooks the vast expanse of Burton Marsh (p. 113), is closed when the range is in use; at such times red flags are flown.

Access to the Steelworks Reserve is by permit only, obtained from The British Steel Corporation. *It is essential that visitors do not stray from the established footpaths whilst on the reserve.*

NUNSMERE

☛ Footpath.

LOCATION
North Cheshire, 2 miles east of Delamere (117) 587688.

ACCESS
From A49(T) Warrington/Whitchurch road. Nunsmere lies along the eastern side of the road approximately half-way between the junctions with the A556(T) and the A54. Parking is in the roadside laybys. A path along the forestry road to the south of the lake can be followed leftwards around the lake, returning to the road leftwards along the disused railway cutting at the northern end of the lake.

HABITAT
Small deep mere and island surrounded by high banks, deciduous and coniferous woodland.

BIRDS
Best known for its regularly occurring **Black-necked Grebes** the mere more commonly attracts **Pochard**, **Tufted Duck** and **Mallard**. While the surrounding mixed-woodland offers many different species including **Blackcap**, **Wood** and **Willow Warblers** in summer, there are fewer year-round residents like the **Jays** and **Green Woodpeckers**, both of which are normally first located by their loud and distinctive calls, the birds themselves being surprisingly as hard to spot as the much smaller **Warblers**.

GENERAL
Although the woodland is Forestry Commission owned the Mere is in private hands and has been under threat of development for watersports which is at present being successfully opposed.

BUDWORTH COMMON COUNTRY PARK & BUDWORTH POOL

☛ Footpaths, Roads, W.C., Dis. (W.C.).

LOCATION
North Cheshire near Oulton Park Motor Racing Circuit. (117) SJ 592654.

ACCESS
From A49 (T) Warrington/Whitchurch road via Coach Road (signposted Oulton Park, ¼ m. south of A49/A54 junction). For Little Budworth Pool turn left at Oulton Park gates and left again at the church for 100 yards. Park on the verge opposite the gate. A public footpath runs part-way alongside the right-hand side of the pool.

HABITAT
Heath, birch and mixed deciduous woodlands, mossy bogs.

BIRDS
In the main wood and heathland species. **Woodcock**, **Redpoll** and **Green Woodpecker** are among the more interesting of the breeding birds. An hour's walk along one of the many paths is bound to produce a good bag of the commoner species, which include the likes of **Nuthatch**, **Treecreeper** and **Willow Warbler**.

The pool is home to **Coot** and **Mallard** with **Pochard** and **Tufted Duck** as regular visitors. Breeding **Mute Swans** are a thing of the past, but with anglers now using lead substitutes for weights, hopefully they could return one day. A bit of 'nicked' groundbait is a small price to pay for their presence.

GENERAL
The Country Park is owned and managed by Cheshire County Council. Fishing on Little Budworth Pool is controlled by Tarporley Anglers' Club.

Nuthatch

HANDBRIDGE CEMETERY

☛ Footpaths, ⇄ (Chester).

LOCATION
Chester (117) SJ 402654.

ACCESS
From Chester ring-road follow signs for North Wales A483. The cemetery is situated immediately on the south side of the bridge crossing the River Dee.

HABITAT
Mature trees and lawns, some shrubbery.

BIRDS
The first thing that strikes the eye is the magnificent collection of specimen conifers interspersed with the more usual cemetery yews. Less obvious are the **Hawfinches** which regularly frequent these trees in winter and several visits may be necessary to locate this regionally scarce bird. More apparent are the **Nuthatches** that find plenty of crevices in the cedar trunks in which to wedge tougher seeds for opening. Summer has commoner species like **Blackcaps**, **Wrens** and **Song Thrushes** breeding in the denser areas while **House Martins**, **Swallows** and **Swifts** often congregate overhead in substantial numbers.

IRONBRIDGE, ECCLESTON

☛ Footpath.

LOCATION
Cheshire, south of Chester. (117) SJ 415626.

ACCESS
From A483 Chester/Whitchurch road roundabout at junction with A5104 Chester/Ruthin road, take minor road (Overleigh road) east to Handbridge then fifth right, signposted Eccleston. Turn left at the 'T' junction 2 miles further on to parking area alongside river.
OR from A483 take a minor road, ½ mile south of A55 (T) roundabout, signposted Eccleston, through village to park as above.

Walk to the right along the river keeping as close to the river as possible to stay on the public footpath. After about ¾ mile, at the 'Crook of Dee', the path turns sharply right through a gate, leaving the river for a short distance before returning to the bank where it stays until reaching Ironbridge.

HABITAT
Riverine walk through farm and parkland with mixed, mainly deciduous woods.

BIRDS
The combined habitat produces a rich selection of birds. **Carrion Crows**, **Rooks** and **Jays** breed and forage throughout the area. The woodlands, particularly around the Crook of Dee, hold many of the smaller species like **Goldcrest**, **Treecreeper** and **Wren**. **Titmice** are well represented including **Marsh Tits**, which are fairly common and the odd **Willow Tit**, which allows a comparison of these two difficult-to-separate species; a short time learning the difference in their calls is a definite advantage. Summer visitors are as varied as the residents; **Spotted Flycatchers** use the fence and gateposts as their hunting-perches, **Yellow Wagtails** inhabit the lawns and grassland whilst the **Willow Warblers** and **Chiff-Chaffs** are appreciative of the insect-rich bankside trees. The river has **Cormorants** and breeding **Mallards** and also attracts the local speciality, **Mandarin Duck**, which have established a small population of five or six pairs around the Eaton Estate.

GENERAL
There is *no* public access to Eaton Hall Estate, which is owned by the Duke of Westminster.

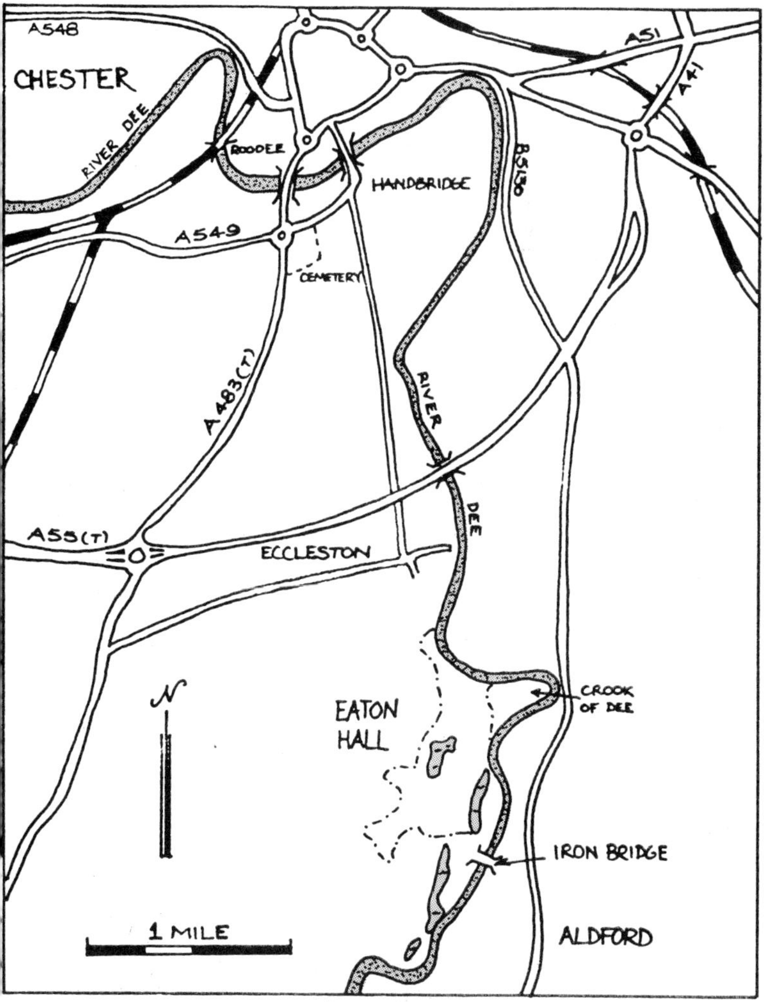

HANDBRIDGE CEMETERY – ECCLESTON IRONBRIDGE

TATTON HALL

☛ Footpaths, Ø, ©, W.C., Dis. (W.C.) Hall & Gardens, (N.T.), Café.

LOCATION
North Cheshire nr. Knutsford (118) SJ 738816.

ACCESS
Leave M56 junction 8 via A556 for Knutsford or M6 junction 19 via A556 for Altrincham. Follow signs from Swan Hotel at Bucklow Hill.

HABITAT
Farmland, parkland, ornamental gardens and lakes.

BIRDS
Since first breeding in Cheshire in 1968, **Ruddy Ducks** have established their stronghold in the county, and Tatton's lakes and meres hold several pairs. Other native wildfowl include **Mallard**, **Pochard**, **Tufted Ducks** and **Teal**; the ornamental ponds in the gardens hold a managed collection of foreign wildfowl which should not be confused with the natural stock. Wood and farmland species abound, among which, worthy of note, are **Woodcock** with their 'roding' flight in spring and **Grasshopper Warblers** which are difficult to see but whose continuous 'reeling' song is unmistakable. Many of the common species can be seen from the system of tracks, paths and nature trails which thread their way around the 1000 acre estate. A summer afternoon's walk should produce **Blackcap**, **Willow**, **Whitethroat**, **Reed** and **Sedge Warblers**; **Blue**, **Great**, **Coal** and **Willow Tits**; **Canada Geese** and the aforementioned ducks; add a few **Corvids**, **Starts** and **Chats**, and the day's list becomes quite a healthy one. Winter is quieter, of course, but **Snipe** move into the boggier areas and an influx of wildfowl appears on the Mere with always the possibility of a **Smew** or one of the rarer **Grebes**.

GENERAL
The Park also contains Tatton Hall which is open to the public and has gardens landscaped by Capability Brown. Fishing is available on the Mere and the parkland holds herds of both Red and Fallow Deer.

The park and gardens, which are managed and financed by

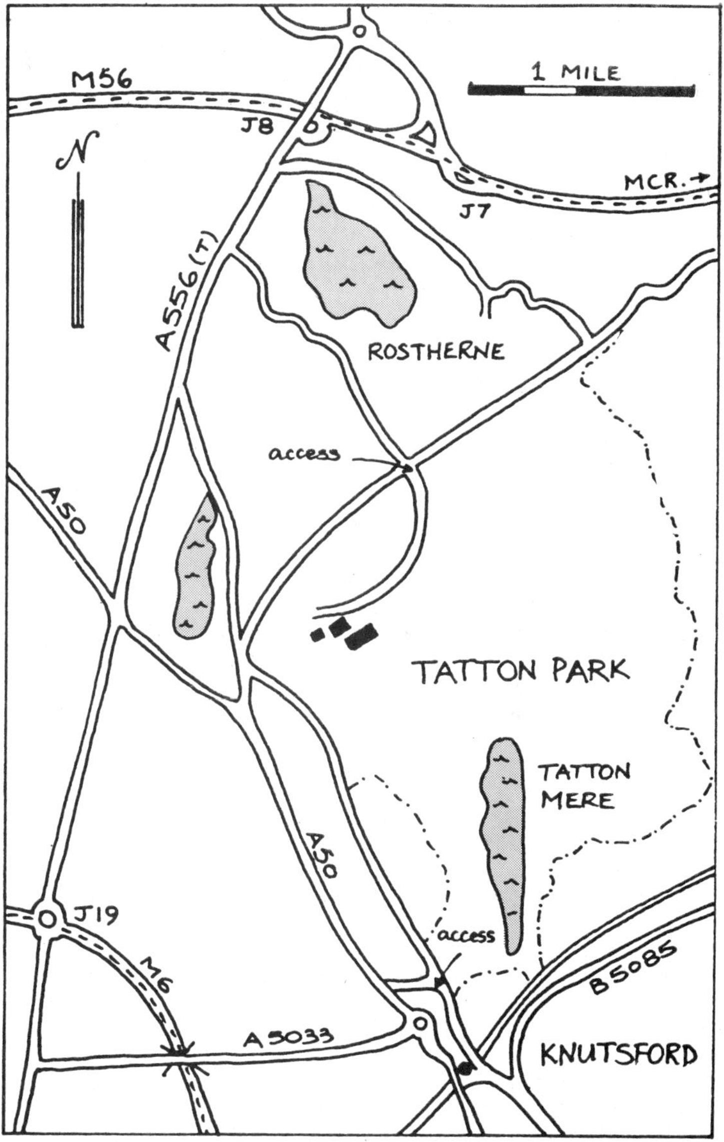
M56
1 MILE
J8
MCR.→
J7
A556(T)
ROSTHERNE
access
A50
TATTON PARK
TATTON
MERE
A50
J19
M6
access
B5085
A5033
KNUTSFORD

TATTON PARK

Cheshire County Council for the National Trust, are open throughout the year, except on Christmas Day and during the Cheshire Show. Times vary, opening between 10.30 & 11.00 a.m. and closing 4.00 to 7.00 p.m., depending on season. Details are available from Knutsford (0565) 54822.

Fulmar

SECTION 4
WIRRAL

CALDY BLACKS

☛ Minor road, footpaths, foreshore. Dis. (Road).

LOCATION
N.W. Wirral (108) SJ 222848.

ACCESS
From A540 Chester/Hoylake Road take B5140 to Caldy turning opposite church into Croft Drive. Follow Croft Drive to right passing cul-de-sac sign to car park on right. Return on foot and follow road to end.

HABITAT
Low sea-cliffs and rocks, estuarine mud and sand.

BIRDS
Perhaps best known for its winter flock of **Scaup**, sometimes into three figures, which occupy the relict channel of the Dee along with **Goldeneye**. Good numbers of **Curlew**, **Turnstone**, **Godwits**, and other waders are also present both on migration and in winter. The resident **Grey Herons** feed along the water's edge at suitable states of the tide when the banks and channels are exposed.

The section of the Wirral Way Footpath which is accessible from this point can produce a good variety of migrant passerines together with resident species and summer visitors of which **Stonechat**, **Whinchat** and **Willow Warbler** are amongst those which breed.

GENERAL
A good place to visit after spending high tide at one of the more notable high tide spots as the channel here holds concentrations of birds at low water. Care must be taken with tide times as the gulleys are deep, as is the mud in places. If in doubt on either point stick to the bank or the sand, although with care, and wellingtons, access is reasonable. Submerged at high water, the Blacks themselves are part of the same sandstone ridge as Tanskey and Hilbre.

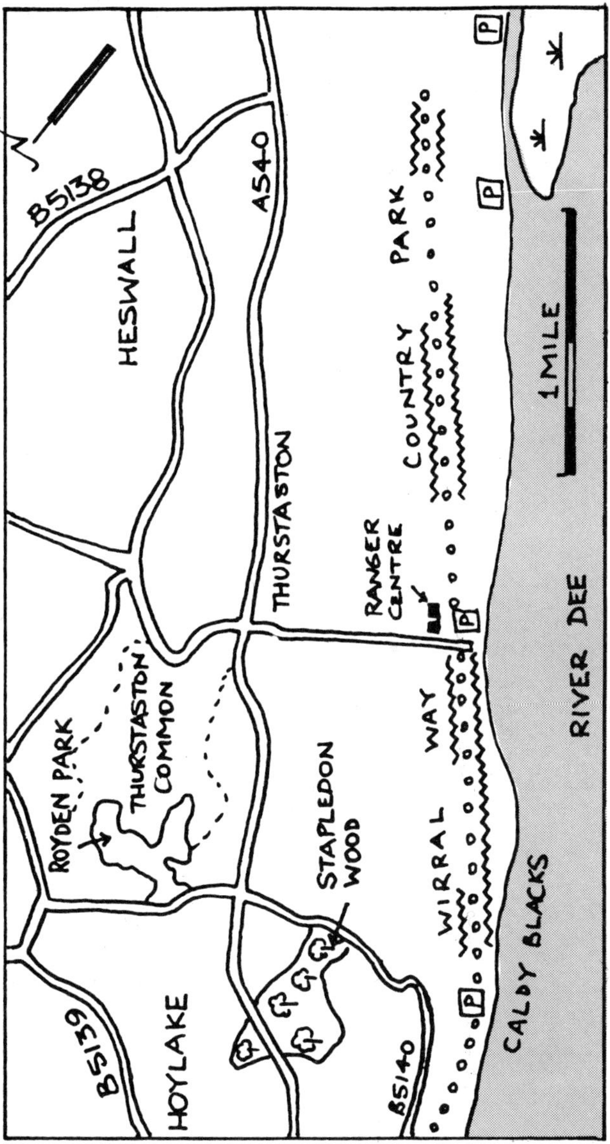

WIRRAL C.P. — CALDY BLACKS — STAPLETON WOOD
THURSTASTON COMMON — ROYDEN PARK

WEST KIRBY MARINE LAKE & TANSKEY ROCKS

☞ Road, shore, Dis. (Lake only), W.C., ⇌ (West Kirby).

LOCATION
N.W. Wirral (108) SJ 210868.

ACCESS
From the A540 Chester/Hoylake Road at West Kirby via Dee Lane to beach and marine lake. Tanskey Rocks lie approximately ½ mile straight out from the marine lake and are accessible by following the ebb tide.

HABITAT
Sandy beach, marine lake, rock and estuarine mud.

BIRDS
Summer disturbance keeps birds away from the marine lake at this time but out of season visitors include regular **Goldeneye** and **Great Crested Grebe**, with the chance of a rarity like a **Red-necked Grebe**. The shore is a haven for **Redshank** and **Shelduck** in the summer; waders abound on migration, part of the massive movement of birds which pass through this area in both spring and autumn. Even when the lake is in use for watersports the estuary can provide interest. The best observation point as high tide approaches is the promenade and sea wall at the south end of the marine lake overlooking the mudflats and small boat moorings, as the incoming water concentrates the birds at this point.

In winter Tanskey Rocks can hold a concentration of birds during low water, sometimes making a good perch for a **Merlin** in search of a meal. The rocks themselves, which although fairly extensive in area are nowhere higher than about two feet (.6m) above the surrounding sand, form part of the same sandstone ridge as the Hilbre group.

GENERAL
The north end of the marine lake is the starting point for crossing on foot to Hilbre (see page 102).

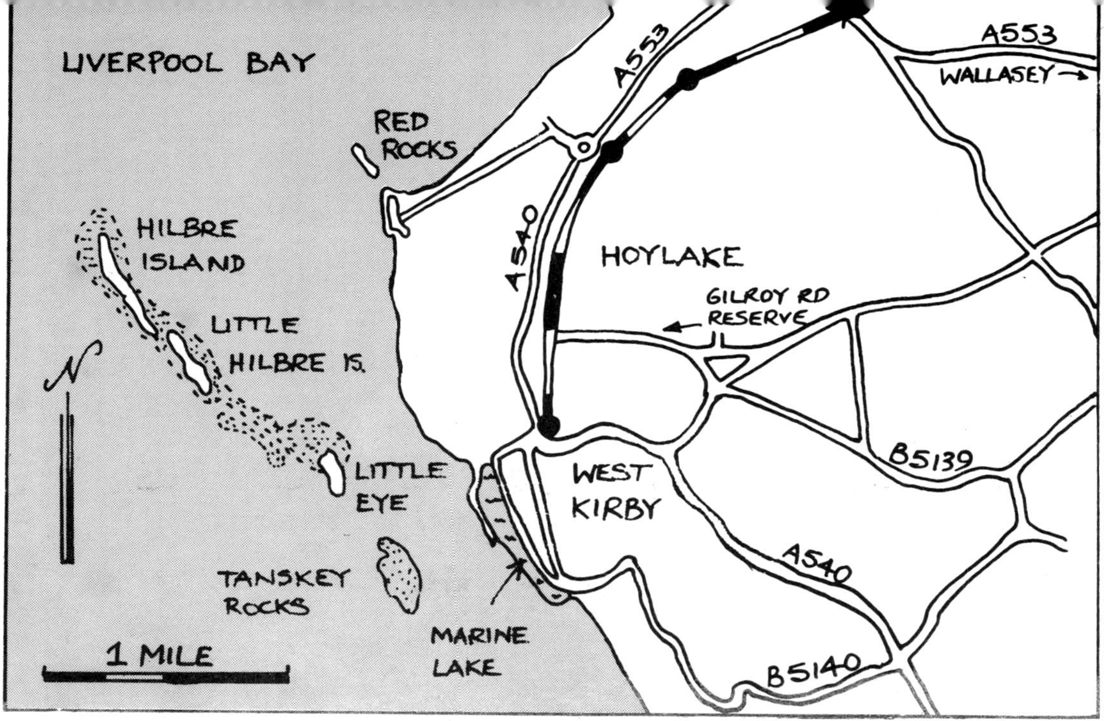

WEST KIRBY MARINE LAKE — TANSKY ROCKS
HILBRE ISLAND — RED ROCKS — GILROY RD RESERVE

HILBRE ISLAND

☛ Ⓡ P, ⇹ S.S.S.I., ⇹ (West Kirby).

LOCATION
Dee estuary 3m. off N.W. Wirral (108) SJ 186877.

ACCESS
A540 West Kirby via Dee Lane, near B5140 junction. Instructions for crossing are on the sign at the end of Dee Lane. Care must be taken to follow the instructions given (see below).

HABITAT
Marine island, sandstone rocks, sea turf with some small shrubs, estuarine mud & sand.

BIRDS
An excellent offshore vantage point for sea-watching gulls and waders, and pelagic **Razorbill**, **Puffin** and **Guillemot** in winter and on passage. Passage migrants include **Common**, **Arctic** and **Roseate Terns**; **Great**, **Arctic** and **Pomarine Skuas**, as well as a variety of passerines which are usually to be found around the shrubbery near the Heligoland traps of the Bird Observatory. Large flocks of **Knot** and **Dunlin** in winter accompany **Grey Plovers**, all of which use the islands as a high tide roost when their feeding grounds are covered. A small flock of **Purple Sandpipers**, which can number around 50 birds at times, is an annual attraction with small numbers also being present in spring and autumn; these confiding birds often stay in the vicinity of the old lifeboat slipway at high water.

As over 200 species have been recorded by the Observatory, the unexpected can turn up at any time.

GENERAL
Do not attempt to cross to the island less than 3 hours before high water and follow the route around the end of Little Eye approaching Middle and Greater Hilbre from the far side in order to avoid the deep channel on the inland side of the islands. There are no toilets or shelter and visitors must be prepared to stay five hours to allow the tide to ebb sufficiently to permit return. Despite these few drawbacks

Black Guillemot

the islands must be considered a prime spot on any list. Generally better if high spring and autumn tides at weekends can be avoided, as there can be more birdwatchers than birds at these times!

PERMITS
Issued free from The Department of Leisure Services, The

Great Skua

Concourse, West Kirby. Individuals can apply for a yearly permit for up to five people. Parties are allocated permits for single visits on a first come, first served basis.

RED ROCKS

☛ Shore, footpath, ⇄. (West Kirby).

LOCATION
Hoylake, N.W. Wirral (108) SJ 203884.

ACCESS
From A540 Hoylake roundabout via The King's Gap, turning left into Stanley Road. Park near gate to beach at end of road. On entering the gate, Red Rocks lie immediately right and the nature reserve is to the left.

HABITAT
Sandstone outcrops on estuarine sand. Reserve area reedbed with coastal scrub.

BIRDS
Winter waders including **Grey Plover**, **Curlew**, **Oystercatcher**, **Dunlin**. Migrants of many species may be found in the reserve, coastal scrub and adjacent gardens. **Red-throated Pipit**, **Broad-billed Sandpiper** and **Hoopoe** are among the many interesting rarities which have been recorded.

Seawatching here is enhanced by the rocky outcrop just off the point, which is surrounded by the sea at high tide, and the roosting birds which settle there are easily viewed from the flat rocks on the shoreline. **Gannets**, **Grebes**, **Auks** and **Divers** are frequently seen offshore outside the breeding season and autumn gales can drive many of the more pelagic species like **Leach's** and **Storm Petrels** within range of the observer.

GENERAL
Accessible at all states of the tide; which makes it a suitable alternative to Hilbre when time or tide do not permit a visit. Do not stay on the offshore part of the rocks as the waves can, and frequently do, cover them, nor will the birds alight in such close proximity to people.

The small reserve provides a breeding ground for the rare and protected Natterjack Toad.

Gannet 2nd year juvenile

WIRRAL COUNTRY PARK

☛ Roads, footpaths, ⇌ (West Kirby, Hooton), Dis. W.C.

LOCATION

Based around 12 miles of disused railway track converted into footpath running from Hooton (South Central Wirral) via Willaston, Neston, Heswall and Caldy to West Kirby (North West Wirral). (108) SJ 237835.

ACCESS

West Kirby and Hooton Stations or any of 17 intermediate access points along the route, all of which may be reached from minor roads off the A540 Chester/Hoylake road. Parking is available at:–

Willaston, Hadlow Road. (station) [B51511]
Neston, Station Road; [via B5136 or B5134]
Parkgate, Old Baths, Boathouse Lane; [via B5134 or B5135]
Heswall, Broad Lane (off Station Road).

HABITAT

Railway cuttings and embankment, gardens, farmland, estuary mudflats and earth cliffs.

BIRDS

For the considerable variety of wildfowl and waders see separate sections on West Kirby Marine Lake, Caldy Blacks, Denhall and Gayton.

The length of this path traverses a wide range of habitat, diverse enough to produce an impressive list of resident and summer visiting species, too numerous to mention in detail. A few **Green Woodpeckers** inhabit suitable sites and are often seen feeding on the ground unlike the **Great** and **Lesser Spotted Woodpeckers** which, although they may also be seen, are more strictly arboreal. **Kestrels** and **Sparrowhawks** are the common predators, but rarer finds like **Hobby** and **Red Kite** have also been recorded. Periodic irruptions of the European population have resulted in sightings of **Waxwing**. **Quail**, probably from birds introduced elsewhere, are being seen progressively more regularly. Good numbers of **White Wagtails** are regular on passage and in winter **Fieldfare** and **Redwings** abound.

The established winter roost of **Hen-Harriers** at Parkgate holds several birds and **Short-eared Owls** join with them in hunting over the Dee marshes.

GENERAL

Visitor and Information Centres have been established at Willaston and Thurstaston, the latter also having a Camp and Caravan site.

Further details may be obtained from The Wirral Country Park Centre, Station Road, Thurstaston, Wirral, L61 0HN. Tel. 051-648-4371.

See also *Walking Northern Railways. Part 2: West* by C. Emett. (Cicerone Press).

BROTHERTON PARK & DIBBINSDALE L.N.R.

☛ Nature Trail, ⇌ (West Kirby), S.S.S.I.

LOCATION
East Wirral between Bromborough and Bebbington (108) SJ 344833.

ACCESS
A41 Birkenhead/Chester Road or M53 (junction 4) via B5137 to bridge in Spittal Road.

HABITAT
Ancient broad-leafed woodland, stream, carr woodland, reedbed.

BIRDS
The varied and well-husbanded habitat available results in the abundance of the many different species that are to be seen here. Many, like **Reed Buntings**, **Reed** and **Sedge Warblers** find suitable breeding conditions as do **Redpolls**, **Jays** and **Lesser-spotted Woodpeckers**.

Kingfishers are often to be seen darting about the stream and **Water Rails** are visitors to the reedbeds in winter, when the resident woodland birds may be joined by visitors like **Siskin** and **Brambling**, feeding on the season's crop from alder and beech.

GENERAL
This small remnant of the ancient Royal Forest, designated as an S.S.S.I., managed and wardened by Wirral Borough Leisure Services, also provides a rich variety of plant, mammal and invertebrate life.

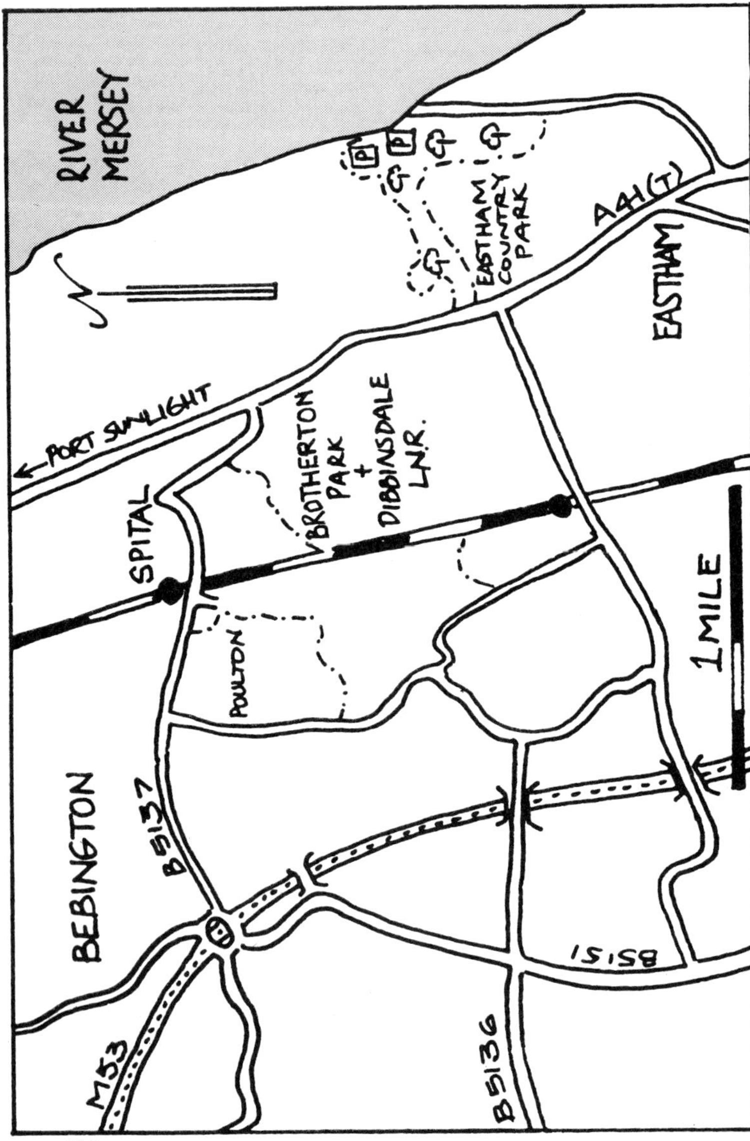

BROTHERTON PARK — DIBBINSDALE — EASTHAM COUNTRY PARK

GAYTON R.S.P.B. RESERVE/PARKGATE

☛, Ⓡ, W, ⇌ (Hooton), Dis. (road), S.S.S.I.

LOCATION
Dee Estuary near Neston (117) SJ 273791.

ACCESS
From A540 Chester/Hoylake road via B5135 to Parkgate, or B5134 to Neston, keeping left in town centre, then on to Parkgate. Parking is available at either end of the Promenade at Parkgate. See also Burton/Denhall and Broken Bank **p.113**.

HABITAT
Inter-tidal mud and sand, extensive saltmarsh with a small reedbed at Neston.

BIRDS
The 5,000 acres of the reserve provide protection for an outstanding variety and quantity of birdlife, particularly Waders and Wildfowl.

Whilst the marsh itself provides breeding habitat for **Redshank**, **Oystercatcher** and **Shelduck**, it is primarily in autumn and winter that the area really comes into its own with the arrival of huge quantities of **Pintail**, **Teal**, **Mallard** and **Wigeon** alongside four-figure numbers of **Knot**, **Dunlin** and **Curlew**. Add a liberal sprinkling of **Grey Plovers** and **Bar-tailed Godwits** and the importance of the area for both birds and the bird-watcher is beyond doubt. The car park next to the Boathouse Inn at the northern end of the seafront is the best spot to watch from, particularly at peak tides (9.8m+) when most of the marsh is covered and **Water Rails** come ashore to hide under the cars and even in the bare branches of hawthorns. Short-tailed voles are also driven out to run the gauntlet of waiting **Herons**, **Short-eared Owls** and **Gulls**, often running over the feet of the static observers to reach long grass at the car park edges. The coastal footpath north can be followed to Hoylake **(see p.105)** whilst the path to the south leads past Neston Reedbed (breeding place of **Grasshopper Warbler**, **Reed Warbler** and **Reed Bunting**) to Burton and Denhall **(see p.113)**.

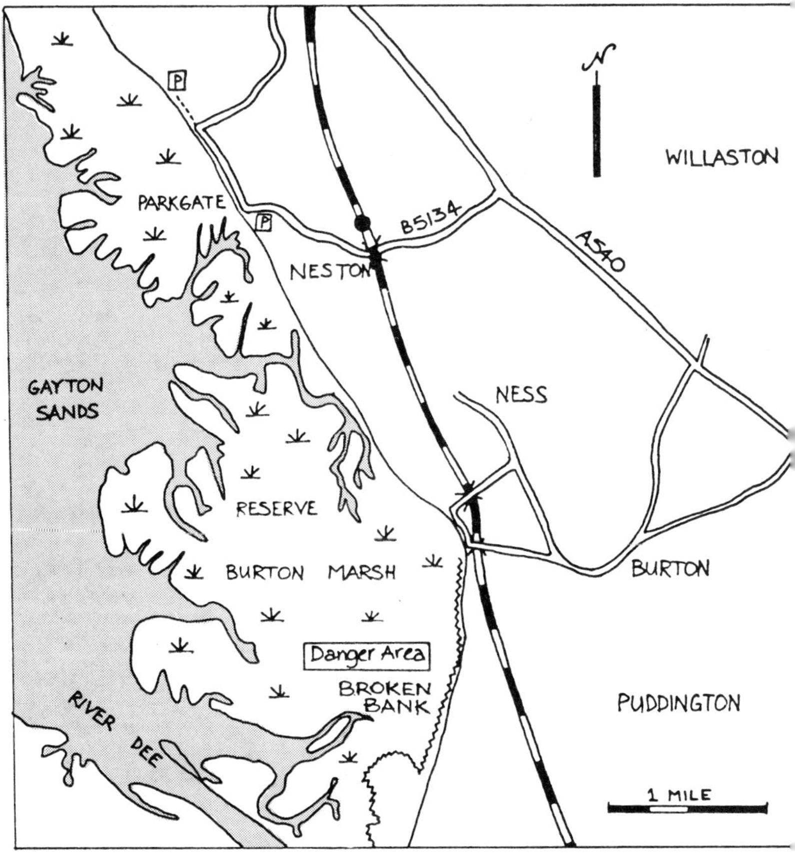

GAYTON SANDS R.S.P.B. RESERVE – PARKGATE BURTON MARSH – BROKEN BANK

GENERAL

Wellingtons or boots are *de rigueur* for attempting the coastal paths in all but the driest weather or following high tides. It is not advisable to walk out onto the marshes as the gulleys and mud are deep and the tides treacherous.

DENHALL, BURTON MARSH AND BROKEN BANK

☛, (part ®, W), Dis. (road), S.S.S.I.

LOCATION
Southwest Wirral at Burton (117) SJ 302746.

ACCESS
From A540 Hoylake/Chester road via Dunstan Lane (signposted Burton) then Station Road *or* Mill Lane (signposted Ness), then Denhall Lane; both roads meet at the side of the marsh. The path to the north leads alongside the southern part of the R.S.P.B. Gayton Sands reserve whilst the minor road to the south (public access on foot only) leads to Broken Bank and Shotton Steelworks (p.84).

HABITAT
Saltmarsh, grazing-marsh and pools.

BIRDS
The path north leads alongside Decca Pools; shallow scrapes which are named after and situated next to the Decca Radio Towers. These pools attract waders and wildfowl including **Mallard**, **Shelduck** and **Teal**, all of which breed on the marsh, but the spot is perhaps better known for the many migrants which pass through here. **Whimbrel** and the various **Sandpipers** are regulars whilst rarer species occur every season.

The main area of the marsh holds plenty of waders and wildfowl in the winter whilst in summer **Linnets**, **Goldfinches** and **Reed Buntings** feed on the proliferation of plant seeds. A pair of **Montagu's Harrier** once summered on the marsh without any proof of breeding. The area has a good population of **Grey Herons** throughout the year and, in the breeding season, whilst the large heronry at Burton Manor is not open to visitors, the birds may easily be seen from the steelworks road as they fly in and out. The surrounding farmland and woodland increases the number of species present by providing alternative habitat and a visit to Burton Woods (N.T.) opposite the Manor can produce all three species of **Woodpeckers** as well as **Jay**, **Nuthatch** and **Treecreeper**. The small pool which is visible in the field next to the road is visited by local **House Martins** as a source of mud for nest-building.

GENERAL

The footpath alongside the marsh continues along the earth bank at the south end of the marsh but this area is often closed when the adjoining firing range is in use. Red flags are flown when the range is in use.

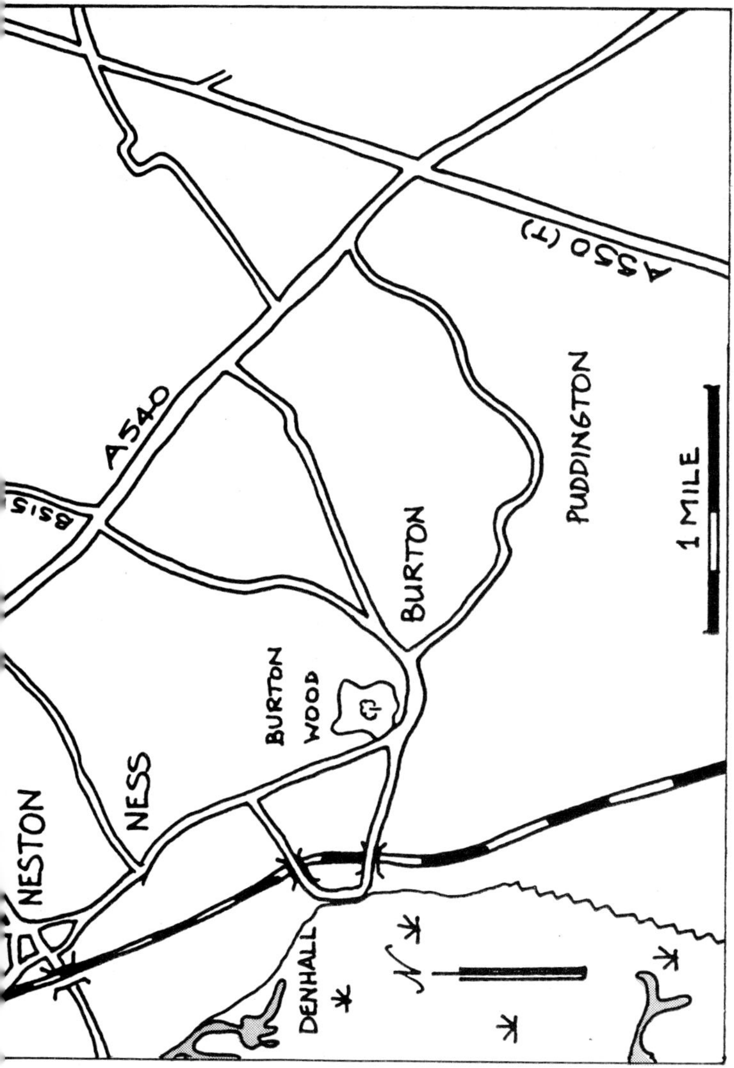

BURTON WOOD — DENHALL — BURTON MARSH

NEW BRIGHTON

☛ promenade, beach, W.C., Dis. (road), ⇌ (New Brighton).

LOCATION
North-east Wirral (108) SJ 313942.

ACCESS
A554 Birkenhead/New Brighton road (Victoria Road) to end of old pier, turning left (King's Parade) or right (Tower Promenade).

HABITAT
Marine lake, inter-tidal sand with sandstone outcrops and mussel-beds.

BIRDS
The marine lake which is situated between the promenade and the causeway leading to Fort Perch Rock often attracts visiting wildfowl like **Goldeneye**, as well as the usual gulls. The breakwaters and sandstone rocks are home to large numbers of **Turnstones** and act as roosts for other waders like winter **Dunlin** and resident **Redshank** which feed on the mud and sand exposed by the receding tide. More unusually, an occasional **Long-tailed Duck** has been seen on the river and a pair of **Dark-bellied Brent Geese** were seen feeding on the sea-lettuce close to the promenade.

The promenade between the pier and Magazine Lane (to the south) overlooks barnacle and mussel covered rocks which attract good numbers of **Purple Sandpipers** in winter. These birds are usually to be found at the outer edge of the rocks when the tide is out and can be difficult to locate from the promenade, so the area is best visited as the tide returns and the birds emerge onto the flatter tops of the rocks, usually in company with a selection of other waders. The famous **Iceland Gull**, "Methusela", which visited the sewage outfall at the bottom of Magazine Lane for more than twenty consecutive winters, is, alas, no more, but other **Iceland Gulls** have been recorded here and **Mediterranean Gulls** are almost common.

GENERAL
Fort Perch Rock which once formed an essential part of the harbour defences now contains a military museum which is open daily (times vary).

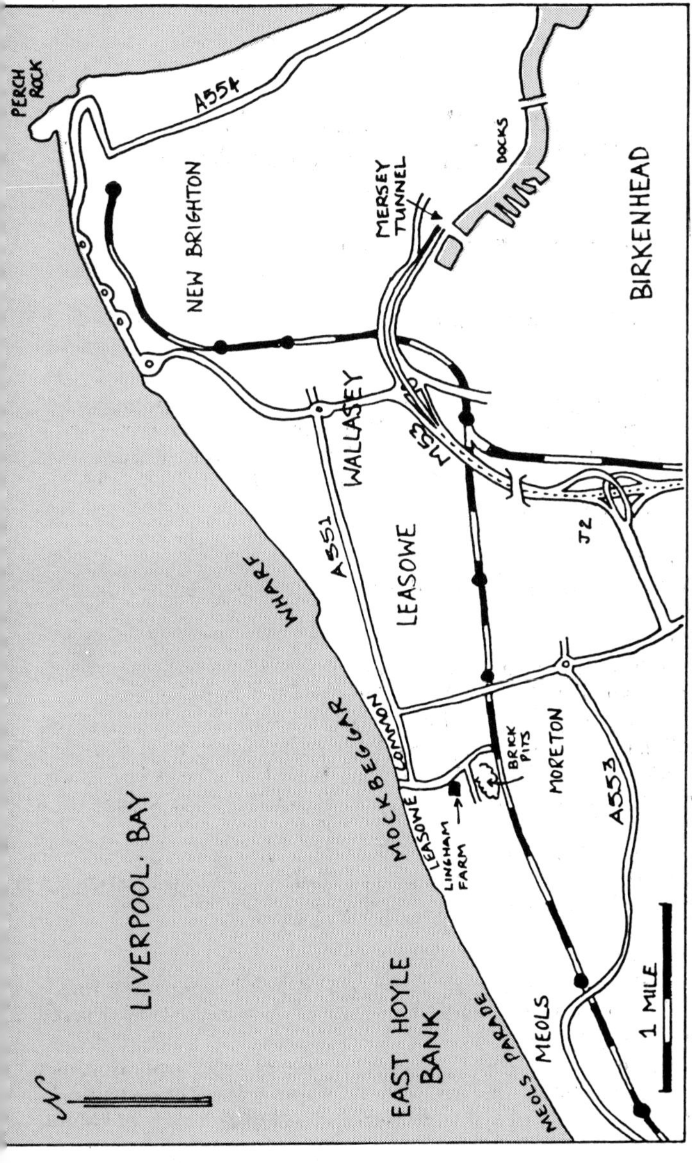

MOCKBEGGAR WHARF — MORETON BRICK PITS
NEW BRIGHTON

ROYDEN PARK AND THURSTASTON COMMON

☛ footpaths, S.S.S.I., ⇌ (Bromborough).

LOCATION
Northwest Wirral between Heswall and Hoylake (108) SJ 243847.

ACCESS
Car park alongside A540 Heswall/Hoylake Road approximately 2 miles (3.5km) north of Heswall. Main entrance to Royden Park is reached from A540 via B5140 to Frankby at the junction with Frankby/Irby road or from B5139 Greasby/Hoylake road then via B5140 to the same junction.

HABITAT
Plantation woodland, parkland with small meres, heath and scrub.

BIRDS
Although good spot for early migrants due to its prominent position, the site has achieved its S.S.S.I. status as a remnant of heathland with a rich indigenous population of flora and fauna. **Woodpeckers** including the locally scarce **Green Woodpecker** inhabit the woodland and park as do **Redpolls**, **Willow Warblers** and **Chiff-chaffs**. **Crossbills** have visited the pines and **Great Grey Shrike** has been recorded.

The heathland and gorse scrub is hunted over by **Kestrels** and **Sparrowhawks** search for unwary **Linnets, Yellowhammers** or **Long-tailed Tits**; all are locally breeding species.

GENERAL
The heathland has areas of 'wet heath' which is interesting for its botany, maintaining plants like Marsh Gentian, Cross-leaved Heather and Sundew.

The common is under active management by Wirral Borough Council Department of Leisure Services to prevent the encroachment of scrub species and to preserve its value as a type of habitat which is in short supply locally.

The visitor centre in Royden Park is open on weekdays from March to October between 2.00 & 4.30 p.m. and on Sundays and Bank Holidays all year from 12.00 to 4.30 p.m.

Great Spotted Woodpecker

EASTHAM COUNTRY PARK

☛, W.C., Dis. P, (W) Nature Trail, ⇌ (Bromborough).

LOCATION
East Wirral near Bromborough (108) SJ 363819.

ACCESS
From A41 Birkenhead/Chester Road at Eastham via Eastham Village Road, then Ferry Road, to riverside car park.

HABITAT
76 acres of ancient woodland and managed gardens bordering estuary.

BIRDS
The nature trail starts off along the riverside cliff top where resident **Herring**, **Lesser Black-backed** and **Black-headed Gulls** are joined, as the tide recedes, by numerous **Redshanks** and **Oystercatchers** while **Shelduck**, **Pintail** and **Tufted Ducks** are among the seasonal visitors. The woodlands contain all three species of **Woodpeckers** along with **Nuthatch**, **Jay** and **Wood pigeon**. Tits abound and, besides the usual **Great**, **Blue** and **Coal Tits**, the ace acrobats of the family, **Long-tailed Tits**, can be seen swinging about the canopy like tiny pink pendulums. **Chaffinches** scrounge for picnic-table scraps and as woods give way to gardens, **Blackbirds** and **Thrushes** are more obvious among the open beds and lawns.

GENERAL
Apart from the birds the area's other wildlife, like hedgehogs, foxes and weasels are principally nocturnal and are best seen early or late in the day.

The Park is owned and managed by Wirral Borough Council Department of Leisure Services.

GILROY ROAD NATURE RESERVE

☛ Footpaths.

LOCATION
West Kirby, North Wirral Peninsula (108) SJ 225875.

ACCESS
From A540 Hoylake/Chester road via B5139 into Saughall Massey Road then Gilroy Road (left). A551 Wallasey/Heswall road via B5139 (as above) at Upton then turning right into Saughall Massey Road at junction with Greenbank Road. Reserve entrance is through large metal gate on the right about halfway down Gilroy Road.

HABITAT
Grassland with scattered shrubs and newly planted trees, surrounding a small Reedmace and rush fringed lake.

BIRDS
A pair of **Canada Geese** have bred successfully over the past few years as have resident **Mallard**, **Coot** and **Moorhen**. The mud at the water's edge provides building material for local **House Martins** and the establishment of blackthorn and hawthorn alongside the path increases the available habitat for hedgerow species. Though small, the reserve attracts a good quantity of species including, on one occasion, a visiting **Bittern**.

GENERAL
Although Council owned, the reserve was reclaimed from a tipping-site by local residents who co-operated in making a local amenity which is actively participated in by neighbouring schools. This far-sighted community approach has created a place of value from former dereliction. The residents operate the reserve under a management agreement with Wirral Borough Council Department of Leisure Services.

BURTON WOODS N.T.

☛ Footpaths.

LOCATION
In the village of Burton, S.W. Wirral (117) 311745.

ACCESS
From A540 Hoylake/Chester road via Dunstan Lane (signposted Burton) parking in village. Access is by the track alongside Barn End Cottage opposite the 'phonebox.

HABITAT
Mature deciduous woodland with small stands of introduced pines.

BIRDS
In summer the woodland is alive with both resident and visiting species like **Chiff-Chaff**, **Nuthatch**, **Treecreeper** and the odd **Wood Warbler**. All three **Woodpeckers** occur and breed, although the hard to find **Lesser-spotted** is more easily located in winter when the leaves are off the trees. **Blackcaps** can be heard calling from the denser corners whilst **Yellowhammers** are often to be found around the fringes and in the adjacent hedgerows.

In winter it is a much quieter place although the more persistent songsters like **Wrens** and **Robins** can be heard and found, and **Jays** can be seen looking for the acorns which they stored in richer times.

GENERAL
Please take particular care in parking so as not to block entrances as roads are narrow and space is restricted. (If combined with a visit to Denhall it's possible to park there and take the short walk back).

MOCKBEGGAR WHARF, MORETON COMMON & BRICKPITS

☛ Footpaths and tracks, seawall and shore, ⇌ (Moreton).

LOCATION
North Wirral Peninsula. (108) SJ 257917.

ACCESS
From A551 Birkenhead/Heswall Road park alongside the track running from right angled bend in main road to Moreton Lighthouse. The seawall path runs west for four miles to Red Rocks (see p.105), via Meols Parade, North Parade and Stanley Road at Hoylake, and east for a similar distance to New Brighton (see p.116). The path to the left just past the Lighthouse crosses the common through Lingham Farm and leads to the Brickpits.

HABITAT
Sandy shoreline, low-lying grass and farmland with banks and hedges, streams and ditches; disused flooded brickpits.

BIRDS
The division between marine and land habitats which is sharply demarcated by the seawall makes the area attractive to a wide range of bird species.

The shore is quiet in summer but, when autumn approaches, the few resident waders are joined by thousands of passage migrants and winter visitors; **Dunlin, Knot, Redshank** and **Sanderling** throng the tideline. Autumn storms often press **Shearwaters** and **Skuas** closer to the land and **Leach's Petrels** can occur in fair numbers during a good blow in September or October. At such times the lee of the seawall offers both the observer and tired migrants welcome shelter from the teeth of the wind.

The common and farmland acts as a 'stop over' for many migrants, often providing the area's first **Wheatears** and **'Chats'** of the year to be seen joining the resident population of **Lapwings** in the search for food. Passerines like **Linnets, Yellowhammers** and **Goldfinches** are common whilst **Grey Herons** fish in the ditches and streams, mostly only to be seen as they lift-off to drift from one to the next. **Mallard**, **Moorhen** and **Coot** frequent and breed around the brickpits near the railway. These are well worth visiting as they can attract less common visiting waterfowl.

Manx Shearwater

STAPLETON WOODS

☛ Footpaths and Tracks.

LOCATION
N.W. Wirral at Caldy (108) SJ 230856.

ACCESS
From B5141/5140 Caldy Road via King's Drive, Caldy, following the road round to the right at rear of Caldy Manor Hospital. Park at end of road.

HABITAT
Mixed and deciduous woodland.

BIRDS
Those seeking after hard-to-find **Lesser-spotted Woodpeckers** may like to try their luck here, because they breed here, but perseverance is required as they are very difficult to locate. This is in no little part due to their small size and habit of keeping to the tops of the trees. It is fortunate that the woods contain many other species, both visiting and resident, which fill in the time spent looking for the **Woodpeckers**. Summertime is best for breeding species like **Redstart**, **Garden** and **Wood Warblers** all of which prefer the broad-leafed woodland habitat.

Adverse weather in spring and autumn can bring a 'fall' of migrants as the wood is a prominently situated refuge. Many fewer species overwinter but those which do are more obvious among the bare branches.

Shoveler

Appendix

USEFUL ADDRESSES

ASSOCIATIONS, GROUPS & CLUBS

Cheshire Ornithological Association,
Mr T. E. Francis, 30, Richmond Crescent,
Vicars Cross, Chester — Tel. (0244) 24236
(See also County Bird Recorders)

Lancashire and Cheshire Fauna Society,
Mr M. J. Butterworth, Hon. Sec.,
7, Hayling Road, Sale, Cheshire — Tel. (061) 962 4007
(See also County Bird Recorders)

Leigh Ornithological Society,
Mr J. D. Wilson,
10 Hoylake Close,
Leigh,
Lancs
WN7 3UG — Tel. (0942) 605627

Merseyside Naturalists' Association,
Eric Hardy,
34, Woodsorrel Road,
Liverpool — Tel. (051) 722 2819

Royal Society for the Protection of Birds,
The Lodge, Sandy,
Bedfordshire,
SG19 2DL — Tel. (0767) 80551

R.S.P.B. N.W. Regional Office,
Imperial Buildings, Imperial Arcade,
Huddersfield,
Yorkshire.
HD1 2BR — Tel. (0484) 536331

R.S.P.B. Local Members' Groups

Chester,	Bernard Wright	Tel. (0829) 40658
Crosby,	Peter Olson	Tel. (051) 924 2744
Liverpool,	Ken Stoba	(051) 428 2350
Southport,	Jackie Rimmer	(0704) 67852
Wigan,	Dr. Geoffrey Ince	(0942) 42113
Wirral,	Kelvin Britten	(051) 653 9751

Woolston Eyes Conservation Group,
Mr B. Ankers,
9, Lynton Gardens, Appleton,
Warrington, Cheshire. — Tel. (051) 926 7355

Wildfowl Trust,
Slimbridge,
Gloucestershire. Tel. (045) 389333

Wirral Bird Club,
Mrs Hilda Truesdale,
8, Park Road, Meols,
Wirral. Tel. (051) 632 2705

COUNTY BIRD RECORDERS

CHESHIRE
Ron Harrison, 8, St. Albans Crescent, West Timperley, Altrincham, Cheshire, WA14 5NY Tel. (061) 928 0047

LANCASHIRE
Maurice Jones, 390, Vicarage Lane, Marton, Blackpool.

Osprey

LOCAL AUTHORITIES, RESERVE MANAGERS & WARDENS

British Steel Corporation,
Shotton Steelworks,
Shotton, Cheshire. Tel. (0244) 812345

Cheshire County Council,
Assistant Countryside Officer,
40A, Church St, Davenham. Tel. (0607) 3874

Croxteth Hall Country Park,
Estate Office,
Croxteth Hall Lane, Liverpool. Tel. (051) 228 5311

Lancashire Trust for Nature Conservation,
Administration Officer,
Dale House, Dale Head,
Slaidburn, Lancashire. Tel. (02006) 294

Haskayne Railway Cutting N.R.

Mere Sands Wood N.R.
Jim Parkinson (Warden),
Rufford, Lancashire. Tel. (051) 6821809

Mull Wood N.R.
(see Croxteth Hall).

Seaforth Dock Nature Reserve,
Royal Seaforth Dock,
Litherland, Merseyside. Tel. (051) 920 3769

"Landlife" Urban Wildlife Unit,
The Old Police Station,
Lark Lane, Liverpool. Tel. (051) 708 5911

Mill Wood L.N.R.

Mersey Valley Partnership,
Camden House, York Place,
Runcorn, Cheshire. Tel. (09285) 73346

Pickerings Pasture N.R.

Merseyside Naturalists' Association,
Eric Hardy,
34, Woodsorrel Road,
Liverpool. Tel. (051) 722 2819

Stanlow N.R.

Nature Conservancy Council,
Blackwell,
Bowness-on-Windermere. Tel. (09662) 5286

Ribble Marshes Reserve Office,
Old Hollow Farm, Banks,
Southport, Merseyside. Tel. (051) 712 5624

Ainsdale Reserve Warden,
2, Westend Lodge, Pinfold Lane,
Ainsdale, Lancashire. Tel. (051) 717 8774

Norton Priory Museum Trust,
Warrington Road,
Runcorn, Cheshire. Tel. (09285) 69895

National Trust,
Regional Information Officer,
Attingham Park, Shrewsbury. Tel. (074377) 343

Formby Point Nature Reserve

Speke Hall,
Custodian, Roy Broadman,
Speke Hall Road,
Liverpool. Tel. (051) 427 7231

Tatton Hall & Country Park,
Knutsford, Cheshire. Tel. (0565) 54822

Pennington Flash Warden Service,
Visitor Centre,
St. Helens Road,
Leigh. Tel. (0942) 605253

Royal Society for the Protection of Birds,
The Lodge, Sandy, Bedfordshire. Tel. (0767) 80551

R.S.P.B. N.W. Regional Office,
Imperial Buildings, Imperial Arcade,
Huddersfield, Yorkshire. Tel. (0484) 536331

Gayton Sands Reserve,
Warden, Colin Wells,
Marsh Cottage, Denhall Lane,
Burton, Wirral. Tel. (051) 336 7681

St. Helens Metropolitan Borough Council,
Dept. of Community Leisure,
St. Helens, Lancashire. Tel. (051) 7424061

Sankey Valley Park,
Visitor Centre. Tel. (051) 7439252

Sefton Metropolitan Borough Council,
Recreation and Amenities,
Town Hall, Southport. Tel. (051) 71 33133

Woolston Eyes Conservation Group,
Mr B. Ankers,
9 Lynton Gardens, Appleton,
Warrington, Cheshire. Tel. (051) 9267355

Wildfowl Trust,
Slimbridge, Gloucestershire. Tel. (045) 389333

Martin Mere,
Burscough, Ormskirk, Lancashire. Tel. (0704) 895181

Wigan Metropolitan Borough Council,
Dept. of Leisure Services,
9, Bridgeman Terrace,
Wigan. Tel. (0942) 44991

Fulmar

NOTES

NOTES

NOTES

NOTES

BIRDS OF MORECAMBE BAY
John Wilson

The warden of Leighton Moss RSBB Reserve tells you where, when, and which birds you can expect to see. From Walney Island to the Lune Estuary.
ISBN 1 85284 000 5
104pp Card cover

FLY HIGH, RUN FREE
Jane Ratcliffe

A leading naturalist tells how she saves injured birds, brings them back to health and teaches them to fend for themselves in their natural environment.
ISBN 0902363 67 0
176pps Card covers

WALKS IN THE SILVERDALE/ ARNSIDE AONB *R.Brian Evans*

A guide to walks and places of interest in this delightful area on the fringe of the Lake District. Numerous illustrations and maps -a guide for the first time visitors and a memento for those who know the area well.
Includes the Leighton Moss Bird Reserve.
ISBN 0902 363 786
160 pps Card covers

WILDLIFE IN MY GARDEN
Jane Ratcliffe

Jane describes how their Lakeland cottage was specifically designed to attract a wide variety of wildlife. Informative and useful. Full colour illustrations
ISBN 0902363 72 7
144pps Card covers

FOUR BOOKS OF INTEREST TO WILDLIFE ENTHUSIASTS

Cicerone Press specialise in books for the outdoors -

Send for catalogue to:-
CICERONE PRESS, 2 Police Square, Milnthorpe, Cumbria LA7 7PY

 PRINTED BY CARNMOR PRINT & DESIGN, LONDON RD., PRESTON, LANCS